Peak Plastic

Peak Plastic

The Rise or Fall of Our Synthetic World

Jack Buffington

An Imprint of ABC-CLIO, LLC
Santa Barbara, California • Denver, Colorado

Library of Congress Cataloging-in-Publication Data

Names: Buffington, Jack, author.
Title: Peak plastic : the rise or fall of our synthetic world / Jack Buffington.
Description: Santa Barbara, California : Praeger, an Imprint of ABC-CLIO, LLC, [2018] | Includes bibliographical references and index.
Identifiers: LCCN 2018028192 (print) | LCCN 2018032207 (ebook) | ISBN 9781440864179 (eBook) | ISBN 9781440864162 (print : acid-free paper)
Subjects: LCSH: Plastic scrap—Environmental aspects.
Classification: LCC TD798 (ebook) | LCC TD798 .B84 2018 (print) | DDC 363.738—dc23
LC record available at https://lccn.loc.gov/2018028192

ISBN: 978-1-4408-6416-2 (print)
978-1-4408-6417-9 (ebook)

23 22 21 20 19 2 3 4 5

This book is also available as an eBook.

Praeger
An Imprint of ABC-CLIO, LLC

ABC-CLIO, LLC
130 Cremona Drive, P.O. Box 1911
Santa Barbara, California 93116-1911
www.abc-clio.com

This book is printed on acid-free paper ∞

Manufactured in the United States of America

Contents

Chapter 1	What Is Peak Plastic?	1
Chapter 2	The Good, Bad, and Unknown of Plastic	15
Chapter 3	The Waste-Driven Supply Chain	41
Chapter 4	2030: A Plastic Tipping Point (Peak Plastic)	57
Chapter 5	The Alternative to Peak Plastic: Exponential Thinking	75
Chapter 6	Solution 1: Stop the Bleeding	87
Chapter 7	Solution 2: Open-Source/Access Plastic (Open-Source Capitalism)	99
Chapter 8	Solution 3: Sustainable Polymerization	111
Chapter 9	Solution 4: A Closed-Loop System for Plastic	121
Chapter 10	Solution 5: Fixing the Invisibility Problem	135
Chapter 11	Summary: Make It Happen!	141

Index 151

CHAPTER ONE

What Is Peak Plastic?

Truth or Consequences?

Science fiction stories always seem to have a purpose. *Godzilla*, the prehistoric sea monster who was awakened and empowered by nuclear radiation, was a metaphor of the impact of nuclear weapons on the Japanese nation, a subject that was still on the minds of its people. *Frankenstein*, written by Mary Shelley in 1831, tells the tale of a creator who has the ability but not the wisdom to re-create life as a lesson for how science during her day was a burgeoning practice of great promise and peril at the same time. Today, I feel as if I must be a science fiction writer telling the story of how the entire planet is being plastic wrapped into submission by the enormous amount of this material that exists on the planet.[1] Is this a metaphor? Hardly so. The story of plastic and its great use in our lives, so seemingly harmless but omnipresent, is a matter few are talking about, yet it is one of the greatest paradoxes we are facing in the 21st century. If you think I'm exaggerating, consider that there is no part of the world that is free of plastic waste—not the deepest part of the ocean, the highest peaks, the most remote areas of the vast blue seas, or the farthest point north at the Arctic Circle. And it is in plain sight of where you are sitting right now, blending into your environment with or without notice, sometimes beyond your vision in particles so small your eyes cannot see. You may not think about it much, but the moment you wake up, your day begins with plastic: as soon as you hit the snooze button on your plastic alarm clock, when your feet slip into your plastic slippers and bathrobe (sometimes cotton, sometimes synthetic, often both), when you pop off the plastic cap of your plastic toothpaste holder (and sometimes putting plastic in your mouth in the form of plastic beads in your toothpaste), wash your face with plastic microbeads in your exfoliate wash from its plastic jar, and then slide over the plastic shower curtain before you pick up the plastic shampoo and body wash containers, and turn on the water that likely consists of undetected plastic nanoparticles unfiltered from the municipal water system. What makes this all possible, intentionally and unintentionally, is a superefficient supply chain system for plastic that serves us as consumers, but also is responsible for this material being recycled only 7 percent of the time. Seventy percent of all plastic ever put into use is now lying in waste in a most unnatural manner somewhere on earth. You will learn from this book that I am a full advocate in the use and growth of plastic in our lives, such as its importance in the health care field, protecting our foods, and nearly uncountable other ways. In a surreal sort of way, plastic is our modern-day superhero, able to defy the laws of nature through being 50 percent lighter than steel at the same strength, leading to half of a Boeing 787 Dreamliner being made from plastic. Also, consider plastic can reside in your body intentionally, such as stents to open blocked arteries, or unintentionally through exposures that are not yet well detected or understood. Maybe even more important is its growing use as a catalysis for good in the developing world, and as a result, plastic use will only continue to grow in use and waste, making it a greater problem in the future than today.

Prior to World War II, the world was a much different place. Since then, annual plastic use has grown from 2 million tons to 380 million tons today, with a worldwide production and use that is hurtling to reach 1 billion tons before the end of the 21st century.[2] No other material, natural or manmade, has ever followed such a trajectory, transforming itself from a mere material into a cultural icon that defines our lives as anything and everything that is easily disposable and replaceable. How can a simple plastic bottle be so useful yet disposable to consumers, irreplaceable as a public service medium to provide fresh water to areas of drought and periled water systems, and yet so nefarious it becomes a market device, allowing companies to charge one to two dollars a bottle for something that can be obtained for pennies from a filtered faucet, and lead to waste? In nature, this seemingly disposable container is anything but so, estimated to persist evilly in decomposition over a period of 500 to 1,000 years in the most unnatural slime path through our ecosystems that nobody really understands. The single bottle in your hand is insignificant until it calculates to billions of them thrown into our environment every year, compounding and decomposing into trillions of pieces of discarded synthetics parsed across the planet in a manner more terrifying than any science fiction novel due to its subtlety. A material validated through science to be so innocuous it can be used in some of our most intimate interactions and yet so destructive in tiny increments of trillions that it could become the most unheralded but greatest destructive threat to our environment just in the period of our lifetime. It's difficult to understand how this is possible, and this book is a journey through the impossible and then the possible. This book is not from my imagination as a science fiction writer, although at times it may seem to be, but rather it is based on data and analysis providing yet another validation to the saying that "truth is often stranger than fiction." I do not believe this is a story of science fiction or conspiracy, but it is true that despite much of the controversy surrounding it, only the plastic resin manufacturers truly understand the bill of materials in its ingredients. A product manufacturer of a cheap plastic toy likely has not much more visibility to the ingredients in its product than does a concerned mother of a toddler who unknowingly chews on it. Likewise, a polymer scientist may understand the differences between polyethylene and polypropylene but only in its base polymer structure and not to its specific formula that can include toxic chemicals either in its design or in its manufacturing. Surprisingly and not surprisingly, there is so little public knowledge about this indispensable material, which is irreplaceable, disposable, and consequential all at the same time.

Maybe the great difficulty in understanding plastic is due to it being considered from the same rules as any other material, such as wood and steel, which are easier to define and describe more discretely. As a matter of fact, wood and steel are easier to be defined as nouns, as things, yet plastic is jumbled in a confusing and complex concept in being a noun, verb, and adjective at the same time. "Plastic" the noun can show up as a material that is nearly invisible and assumed to vanish into thin air (when it does not) like glitter and as a polyvinyl chloride (PVC) pipe that can last for a 100 years or longer. The disposable Styrofoam cooler that crunches when it breaks is plastic, as is also the stretchy polyester shirt you wear; the list of its variations can go on and on. Plastic is *not even plastic*. That is, how we define plastic as its base polymers are often modified through the use of additives, further complicating the material, its supply chain, and its ethos. Its mystique has even led it to gaining a cultural status of being synonymous with the term "fake" and able to be used to depict our 21st century society as shallow, lifeless, disposable, and cluttered. Plastic's very name originated from the Greek word *plastikos*, meaning "to mold" or "form," further defining it better as a verb than a noun as a result of its special quality in being able to become anything we want it to be. It has become a replacement for nature itself; it can be a fake plastic tree in your office or replace the grass field you played on that your kid now plays on as synthetic grass. Our society is being defined with plastic, both literally and figuratively, and as a result, nature is being filled with it as well, leading to unintended consequences.

Because so few of us are talking about this problem is more reason why some will believe this is a story of science fiction rather than fact, a great threat that has gone unnoticed right before our eyes. Some may contend this as a great conspiracy lodged against citizens by Big Business, or others may consider it is as one great fraud exaggerated by environmentalists being no threat other than a few misplaced pieces of trash on their favorite beach.

How many of us would pay attention if a breaking news headline on CNN noted the one trillionth piece of plastic entered into the ocean this year or the Great Pacific Garbage Patch was no longer thought to be the size of Texas but now three times larger? News stories and sites get watches and clicks from terrorism attacks and threats that are so rare your probability of being affected is 1:45,808 versus a 94 percent likelihood your drinking water will contain unwanted traces of plastic.[3,4] This nonchalance from the population regarding the problems of plastic is largely due to a common myth our plastic problem can be fully resolved simply if we as consumers were willing to recycle more—an urban legend that is far from the truth. When it first entered the marketplace, plastic was actually an environmentally sustainable option in comparison to an overbearing dependency on rubber trees, tortoise shells, elephant tusks, bug secretions, and other natural resources threatened to be harvested into extinction. During World War II, plastic became even more important as a strategic advantage to win the war, an arsenal for democracy. After the war, a growing concern emerged of the economy falling back into the ditch of a depression, and plastic became a new strategic imperative to create a consumer-based economy that enabled growth and prosperity. Today, plastic remains as an economic driver; more than you may believe, that is a critical, unheralded element of our conventional economy. Plastic's favorable position in the economy, however, cannot be substantiated through mathematical analysis of how trillions of tons of plastic over a century can be averted from the environment or impacting our own health through a recycling program or other mitigation schemes. From a supply standpoint, there is just too much of it with too little value after its use. This is the paradox I am calling "Peak Plastic," the upcoming point we will reach when the material's use hits a threshold when its marginal benefit to society is less than its cost to the environment. By my estimates, the world will reach this point of Peak Plastic by 2030, although in some places in the world, a case can be made that we are reaching it today or have already reached it.

There is an assumption that plastic as a material is inert, and it probably is in low volume, and contained uses, yet it is also resistant to nature, especially in the scale it is being used today. In nature's supply chain, there is no such material as plastic, meaning there are now organic and synthetic polymers of two different worlds. An organic polymer has been defined by evolution and grown through a closed-loop system where the waste from one stream becomes an input material into another. In this model, materials are synergistic within the ecosystem, thriving in a self-regulating, efficient system. Plastic is neither synergistic nor self-regulating, and it thrives as an evasive synthetic species to these ancient rules of biology. No later than 2030, a tipping point will be reached when the large, compounding volume of this foreign material is no longer tolerable to the point it creates both an environmental and economic crisis combined. This is when the last plastic bottle on the beach is not so harmless, or simply an eyesore.

You do not need to be a biologist to understand that nature wins as a self-regulating system, and it is humans as a species who will lose, not the planet, per se. Certainly, man is the great inventor of this plastic material. We have leveraged the geological processes that fermented dead leaves and animals into energy over tens of millions of years and converted it into hundreds of millions of plastic bottles and other disposable articles annually. Ironically, plastic is economical in use for 15 minutes, only to be crassly thrown away for hundreds of years to leach unnaturally into every nook and cranny on the planet. Nature will adapt to this threat and the mass infiltration of additive chemicals left to leach uncontrollably through trillions of individual articles into the oceans, rivers, lakes, groundwater, land, wildlife, and our bodies. But what will be the result of this, the result of us, as these trillions of *nature resistant* pieces of plastic tear down into micro- and nanoparticles from various forms, including facial scrubs, clothes, dishwashing pods, road tires, and other items unintentionally released into our water systems daily? I am not speaking in metaphors or exaggerating in making these points; I will use this book to ask these questions in order to form answers.

This problem of our apathy seems to be due to us believing we have the solutions to the plastic problem and we just need to follow through on them. Research this question and you are sure to find many experts who profess we can lick the plastic problem simply by consumers doing their part and recycling this plastic, implying businesses are waiting for these materials to come in and that's the only drawback preventing a closed-loop supply chain process. In some sense, this used to be the case when nations like the United States would recycle and import its used plastic to China for its sorting and use, but today, China no longer wants our trash. As I will explain in this book, and did already in my last book, *The Recycling Myth*, believing these activities will lead to salvation is a dangerous misconception and may actually be unintentionally preventing the right solutions to be enacted. Another false narrative not proven through research is to replace the fossil fuel feedstock of plastic with some form of plant-based or bio feedstock to remedy the problem. Such alternatives are no different from fossil fuels–based plastic in an end-to-end supply chain analysis that takes into consideration the polymerization, manufacturing, and full supply chain process of plastic. In this book, I will not take shortcuts, and the plastic paradox will be understood and ultimately supported from an overall supply chain perspective.

There is no question that after a 100 years of plastic use we've really dug ourselves a hole related to this problem: this massive and efficient supply chain system for markets has led to 8.3 billion tons of plastic waste that has flowed into our natural ecosystems—an amount equal to 25,000 Empire

State Building skyscrapers but dispersed in tiny fragments unnaturally across the planet from the deepest reaches of the ocean to the highest peaks and from the most to the least inhabited places, with little regard or understanding.[5] A simple plastic bottle, an amazing supply chain innovation in itself, weighs less than 13 grams through dematerialization that has led to a reduction of half a million tons of PET plastic, reducing the freight costs to ship the bottle to retail.[6] An innovation of this magnitude, impossible to conceive just a mere few generations ago, now has become a marketing marvel and, at the same time, a planetary crisis. A harmless piece of plastic perhaps, but in the words of the famous 15th-century physician Paracelsus, "the dose makes the poison," meaning in totality, it may be enormously toxic. A market solution achieved by the supply chain system is leading to an environmental problem, increasing the costs and benefits in a concurrent manner. While this may seem to be a paradox of sorts, it can be addressed without taking away the benefit while at the same time curing and enhancing the environment. It just needs to be a reality we see differently in our minds and solve it accordingly.

What is plastic and how should it be defined? According to the International Union of Pure and Applied Chemistry (IUPAC), plastic is a polymeric material that may contain other substances to improve performance and/or reduce cost; it's almost always synthetic, most commonly derived from petrochemicals, and exhibits high molecular mass and plasticity.[7] This is a sterile, scientific definition useful to a polymer scientist but not very useful to those of us wishing to understand what is certainly a supply chain and cultural phenomenon in how it is consumed and wasted as a disconnection between society and the environment. In this book, I will provide new meanings for plastic that connects us to what it has become to us and our society, both the good and the bad. The United Nations has defined plastic waste as a planetary crisis, an emotional classification of pathos and an absence of the ethos and logos necessary to solve the problem. No doubt this book will contain some emotional elements, but it will be unintentional as a circumstance from truths so shocking. In this book, I hope to take you on a sober march of how, if we do not act, we will reach this concept of Peak Plastic and the very difficult circumstances as a result. The purpose of a message of how to fix a problem in the next 12 years is less to evoke emotion and more to prepare a road map for action to prevent it from happening, leading to a synergistic connection between our natural and synthetic worlds. The purpose is to change the narrative, as to paraphrase the old saying, "If we change the way we look at things, the things we look at change."

Prepare yourself that in this book there are no simple answers to fix the problem, such as to stop using plastic, to recycle, or to replace it with some organic feedstock; if any of these ideas were effective, we would be on the righteous path of fixing it. Instead, the variables within this topic are often contradictory and paradoxical, such as whether recycling leads to good or bad behavior, whether our health is improved or worsened in the use of plastic, and whether developing nations should be able to grow in use of synthetic materials without adequate waste-management systems. Scientific discovery is a critical component to the solution but not sufficient; it must become an element of the overall supply chain system solution. I will speak the language of supply chain management in this book, the same language that led to the growth in our use of plastic from 2 million to almost 400 million tons and will take us to 1 billion in the future! The math of this problem requires business intervention: a 9 percent annual growth rate that is recycled at less than a 10 percent rate is an environmental catastrophe; the recycling and reuse rate needs to be 90–95 percent, as I will propose in the solutions of the book. When we study this paradox from a supply chain perspective, solutions will not only become possible but also become probable through market activities rather than costly, ineffective legislative schemes that require the consumer to be punished for the bad design they did not create. The innovations I will propose are neither gimmicks nor conceptual academic exercises but principles founded on supply chain solutions that were the foundation for how plastic grew in the first place.

What Is Peak Plastic?

To a hardcore biologist, anthropogenic activity is just another biological routine where our species' progress will lead to its own demise, and man is classified no differently from any other. I am not a trained biologist, yet I cannot agree with such a fatalist point of view that believes we cannot stop ourselves before going over the cliff's edge or have an inability to measure a cost to its related benefit. And yet in defense of the biologist, we are hurtling ourselves toward Peak Plastic; this tipping point, when reached, will have subsequent costs that overrun the benefit. I believe this to be more of a blind spot that can be fixed than an impending doom. An example of this is our lack of visibility of what plastic is and is not. Today, the only plastic classification available to the consumer is something called the resin identification system (RIC) or, as you may know of it, the chasing arrows with the number inside of them. The RIC designates plastics into one of seven categories of design with the seventh being a catchall that includes everything not in one of the first six. This categorization system was enacted in 1988 to assist burgeoning recycling programs in the United States to help bring awareness to consumers regarding what is recyclable and what is not. It is very simplistic because it was never intended to educate consumers to the ingredients of plastic but rather to recycle a material never designed to be recycled in the first place. In a limited sense, the RIC system has been helpful to consumers in understanding different types of plastic if they care to do so even though it has done little to improve the United States' moribund 7 percent overall plastic recycling rate.

And this becomes the problem: should industry create a classification system to educate the public regarding plastic if nobody seems to care? It becomes a vicious cycle for how apathy leads to a lack of education, and a lack of education leads to an insurmountable problem. The plastic resin manufacturers and the petrochemical industry have little incentive to voluntarily challenge their own status quo, and there is little to no potential for government legislation to accomplish anything given the track record of failure of the Environmental Protection Agency (EPA) to enforce existing policy, much less new rules. This leads to consumers who are a bad combination of being uninformed and apathetic, inattentive to a mega-scaled plastic supply chain system growing leaps and bounds worldwide without sufficient checks and balances and spanning numerous different municipalities with government institutions powerless to contain it. Peak Plastic is a matter of too much plastic in our ecosystems, which is certainly the case, but also of not enough information in the public forum.

Can we better classify plastic appropriately if it is not considered a material but rather a system? Worldwide, the plastic supply chain system achieved a compound growth rate of 8.4 percent from 1950 to 2015, with recycling/reuse rates declining and a poor definition of this system to even begin to address the problem. During this era since 1975, global resin production has accelerated by 620 percent, the same year a member of the Council of British Plastic Federation and fellow of the Plastics Institute remarked, "Plastic litter is a very small proportion of all litter and causes no harm to the environment except as an eyesore."[6,8] Today, we consider plastic to be a material when, in fact, the real problem seems to be the system that has created, produces, and distributes it. Yet, when we look at it as a system from an overall supply chain that includes the environment, it falls short; 88 percent of the planet today is polluted with plastic waste, a problem far beyond being an inconvenient eyesore.[9] Back in 1975, polymer science was a brand-new field, and a decade later the field of supply chain management came into its own, resulting in the two disciplines merging forces to enable this supply chain of plastic. Since then, approximately half of the plastic waste that exists in the environment has occurred given a global rise in production and a recycling/reuse rate of less than 10 percent. This is not a coincidence, and systems matter in how science is applied through supply chain, with supply chain becoming the problem through the environment: a perfect storm of polymer scientists working alongside supply chain experts who supersized plastic, changing the metaphor of plastic from being a harmless material.

Peak Plastic is not a metaphor but a line in the sand by which to reinvent the entire plastic supply chain system to ensure its future benefits don't exceed its costs, and to increase its value, enabling plastic to become something that improves the environment through our use. That's an obviously ambitious agenda, and in the next three chapters I will make a case as to why this is so crucial, consistent to the United Nation's belief that we have reached a "planetary crisis" *today*, not 2030. From this sense of urgency, I will provide five solutions for how to combat the problem, starting with an immediate stopgap to today's massive imbalance in order to buy some time for transformation solutions to take form. These temporary solutions must immediately address a tipping point already reached in our oceans and coastal slums of the developing world, and even in our own homes where we eat over 100 microfibers each meal largely from dust that settles onto our plates.[10]

I will make a case for science and supply chain to be linked to our natural ecosystems, becoming a combined supply chain system between industry and nature. Through these connections, I am hopeful plastic, as a material, can transform from being bad for the environment and good for society to being good for both at the same time. However, if we ignore the signs and do nothing, I will discuss how this tipping point will be reached by 2030, where plastic, as the material in the system, is no longer beneficial to the economy or environment, therefore becoming bad for both. We actually have a sneak preview of sorts to this problem through what's happening already in the slums of Asia: to see its debilitating harm, with much of why it is happening related to the global supply chain system that serves us with many of the products we purchase at a Walmart store on a daily basis. Even with our waste-management systems being generally up to the task in the developed world, we are reaching a saturation point where plastic is strewn across our towns and coastlines, as is the case in Asia, and will be insidiously present but not visible in our meals, drink, and other intimate encounters. We are just now coming to grips with the threat as it exists.

The year 2030 is not intended to be a great prognostication of any sort but rather a guidepost for us to use to put in place temporary, immediate, and long-term sustainable solutions to reverse, correct, and improve this problem for the future. In areas of such fast-growing nations of Asia as China, what's happening seems to indicate a more immediate action plan needed than what seems proper for such nations as the United States, but this is consistent to how the global supply chains operate that should be better understood. And not just in the global supply chain but also the supply chains of our environment, where damage to our oceans and rivers in these places does not just impact their food chain systems and health but ours as well, thousands of miles away. If there is any good news in the story, and there is, it has never been a better time to address how these critical fields of polymer science and supply chain systems can *do good* in a global system rather than leading to economic and environmental conflict, as exists today.

What This Book Is About

A false dichotomy exists today wherein some consider plastic to be purposely good, while others believe it to be wickedly bad. Both these views cannot be correct as definitive statements, and the truth is neither of them are; plastic provides both a net benefit to society we cannot do without while also becoming a growing existential threat we cannot live with either. The purpose of this book is not to be able to answer all these questions but rather to validate them as legitimate concerns and to put forth the appropriate degree of concern around solving them. To identify a problem of this level of mystery will require a multidisciplinary approach rather than taking on this complex problem from only one angle. In the next chapter, I will define plastic in an unconventional, multidisciplinary manner by weaving together polymer science, supply chain, public policy, environmental science, and other fields to demonstrate why plastic is more than just a material and can be considered a culture just the same. Might plastic be harmful to our health? It is a question that must be considered without being burdened as to whether it is correlative or causal. Concerns appear to be inconclusive at this point, but that just means we do not know, and the scale and complexity of plastic material makes it difficult to discern. In chapter 2, I seek to weave a blanket of the many dimensions of plastic without an indisputable verdict, but opening the possibilities allows us to change its metaphor; this degree of understanding must be much deeper than that of plastic as a material.

After this baseline of a technical, supply chain, and policy discussion of plastic, chapter 3 delves into how these materials can impact both our society and natural ecosystems, with a primary focus on the oceans. Our oceans are a source of food and oxygen critical to our health and safety. There is sufficient evidence our oceans are on the brink of collapse through this mass introduction of synthetic substances, mainly plastic, its associated chemicals, and carbon dioxide as a matter of climate change. Certainly the debate on climate change is a contentious and political one, at least in the United States, and yet there are solutions in motion to address them, such as electric cars and renewable energy. In contrast, plastic systems have few meaningful alternatives for how it will stop harming our oceans and potentially our own bodies. In regions of Southeast Asia, a tipping point is happening today, as I will illustrate in this chapter. After this overview of plastic and its impact to the environment, chapter 4 addresses the concept of this Peak Plastic tipping point and what it would mean if we reach this point. If this tipping point indeed matters, what is its impact to us, and why should we care? This is an important chapter, since most Americans and others in the developed world pay little attention to this seemingly harmless material and what impact it may have to our future.

For a problem of this magnitude, where do we start? Chapter 5 is the commencement of how technological and supply chain innovation can mitigate and eventually solve for this plastic problem in industry and nature. In this book, the epicenter of solving the problem is through innovation of the end-to-end supply chain system rather than isolated developments in material science or public policy/regulatory options, which are not working today and have not worked in the past. Reinventing uses of materials may become a part of an overall supply chain solution, as may some forms of regulation and even bans (e.g., microplastics), but it must be taken on as an overall holistic approach rather than individual initiatives. When I studied this topic in Sweden for my PhD studies, I found their approach to the use of culture as a very important foundation, but what they lack is an end-to-end structure to really solving rather than mitigating it. If we can take the culture of the Swedes and the supply chain innovation of the Americans, we can solve this problem from cultural, economic and environmental standpoints rather than one versus the other. With a focus on the right problem, the right culture, and the right tool (the supply chain system and emerging technologies), there are great opportunities to link industry synergistically to the natural world. The use of the supply chain system to solve the plastic problem it created seems to be a logical path of addressing this paradox by supply chain professionals, and this idea will be introduced in this chapter of the book. The five solutions I recommend, based on innovation, will provide a renewed sense of what is possible in the next decade when we change the supply chain system rather than the material; plastic need not disappear from our world, except from the natural environment.

Chapters 6 through 10 are the five solutions for creating a paradigm shift of how plastic transforms from a disruptive to enabling force in our economy and environment. Chapter 6 is the Band-Aid that will hopefully hold until more innovative, sustainable solutions can be put into practice. In chapter 7, I introduce a new design and innovation system that can disruptively transform our view of plastic as this private enterprise where we participate only as passive consumers, to us driving the change as both consumer and designer through our understanding of the challenge. Through this momentum, the market will ferret the true identities of these plastics in all their forms rather than waiting for something to happen by regulators, or even industry without sufficient market pressure. The key to achieving supply chain transformation will be an open-source, open-access model rather than proprietary methods. Now that design through waste is in process of transformation, chapter 8 will highlight the greatest challenge of plastic in our natural world, that of industries' clunky approach of polymerizing, depolymerizing, and re-polymerizing much different from nature. As I've mentioned, plastic as a material is not really the problem, as much as the plastic development process. Once this has been addressed, chapter 9 will focus on the reality of a near 100 percent closed-loop system like exists in nature. This term "closed-loop system" is bandied about liberally in the media without a true definition because it is not comparable to nature; it should not be used. To the contrary, if these solutions proposed in this book are enacted, a closed-loop system as exists in nature will happen. The fifth solution, and probably the most difficult, is to get our arms around the plastic supply chain we cannot see, which presents the most potential harm to us and the environment. The problem, of course, is to first detect before anything can be solved, which is the greatest challenge of all. After these solution chapters, I will close the book with the math and a road map of how to make this possible.

Sometimes when I think about this all-encompassing role plastic has taken in our modern lives, I wonder if I am living in some kind of dream given the banality of something so good being potentially so bad before we get to understand exactly what plastic is and is not. As time passes, the dream is turning into a nightmare, especially when we travel to one of many developing megacities (especially near an ocean) and see it for ourselves. Plastic is not a material, or even just an eyesore, but rather a system upon itself and separate from nature until it becomes waste, and then it is inextricably linked as an ecological and potentially financial and health problem. Truth be told, we really do not know what it is; we don't have enough information to know, and it's too bad so little knowledge has been gained over the past decades to have a better understanding than we do today. So let's accept where we are today, with no more time to wait, and let's move forward and develop an industrial model that connects us and itself to nature.

References

1. McKie, Robin. (January 23, 2016). Plastic now pollutes every corner of Earth. *The Guardian*. Found at: https://www.theguardian.com/environment/2016/jan/24/plastic-new-epoch-human-damage (accessed June 13, 2018).
2. Dengler, Roni. (July 19, 2017). Humans have made 8.3 billion tons of plastic. Where does it all go? *Science*. Found at: https://www.pbs.org/newshour/science/humans-made-8-3-billion-tons-plastic-go (accessed June 13, 2018).
3. Mosher, Dave, and Skye Gould. (January 31, 2017). How likely are foreign terrorists to kill Americans? The odds may surprise you. *Business Insider*. Found at: http://www.businessinsider.com/death-risk-statistics-terrorism-disease-accidents-2017-1 (accessed June 13, 2018).
4. Boesler, Matthew. (July 12, 2013). You are paying 300 times more for bottled water than tap water. *Slate Magazine*. Found at: http://www.slate.com/blogs/business_insider/2013/07/12/cost_of_bottled_water_vs_tap_water_the_difference_will_shock_you.html (accessed June 13, 2018).

5. Glowatz, Elana. (July 19, 2017). Humans have produced enough plastic to build 25,000 Empire State Buildings. *International Business Times*. Found at: http://www.ibtimes.com/humans-have-produced-enough-plastic-build-25000-empire-state-buildings-2568044 (accessed June 13, 2018).
6. Lauria, Tom. (February 18, 2010). Weight of PET bottled water containers has decreased 32.6% over past eight years, saving 1.3 billion pounds of plastic resin. *International Bottled Water Association*. Found at: http://www.bottledwater.org/news/weight-pet-bottled-water-containers-has-decreased-326-over-past-eight-years (accessed June 13, 2018).
7. International Union of Pure and Applied Chemistry. What are polymers? Found at: https://iupac.org/polymer-edu/what-are-polymers/ (accessed June 13, 2018).
8. Moore, Charles J. (October 1, 2008). Synthetic polymers in the marine environment: A rapidly increasing, long-term threat. *Environmental Research* 108 (2): 131–139.
9. Warlia, Arjun. (July 7, 2014). New study finds 88 percent of Earth's ocean surface now polluted with plastic trash. *Collective Evolution*. Found at: http://www.collective-evolution.com/2014/07/07/new-study-finds-88-percent-of-earths-ocean-surface-now-polluted-with-plastic-trash/ (accessed June 13, 2018).
10. Rosane, Olivia. (April 5, 2018). Humans eat more than 100 plastic fibers with each meal. *EcoWatch*. Found at: https://www.ecowatch.com/plastic-consumption-food-dust-2556502607.html (accessed June 13, 2018).

CHAPTER TWO

The Good, Bad, and Unknown of Plastic

In the Beginning

The story of plastic begins long ago, tens of millions of years before humans entered the scene, across the oceans and land. Fossil fuel, as the name suggests, is an energy source made from dead animals and plants, an energy source of a consistent molecular structure with living things today. Scientists have pieced together the story of how, over an epoch of millions of years, these organisms were transformed into potential energy after being buried for so long in the sediment and converted through a proper combination of temperature and pressure into sludge that today is being used, among other things, to make plastics. Through a benefit of geological time, heavy molecules called kerogens were catalyzed through high temperatures and pressure, causing these molecules to crack (complex organic molecules breaking into simpler ones) and allowing them to become a highly concentrated source of energy. This has been the most powerful, efficient form of energy accessible to man, able to be burned and transformed into matter for our industrial supply chain. This discovery of dead organic matter that can be transformed so brilliantly for our use is a remarkable discovery, yet it is no match for nature, which remains the better material scientist for having created energy and material through a closed-loop system that transforms waste back into an input, a process that has been perfected over hundreds of millions, if not billions, of years of evolution. In nature, material science is a self-ordered structure developed through time by the utilization of the same properties of atomic hierarchy we seek to copy through much simpler manmade constructions.[1] Humans as material scientists have come a long way in the development of crude forms of metals, polymers, composites, ceramics, and glass that have historically been used for our societal benefit.[2] Over the course of history, material innovations have been so important in defining our periods of time and our progress, such as the Stone, Bronze, and Iron Ages. Man started by manipulating the natural polymers of wood, animal skins, and fibers, before moving to metals and then to the synthetic polymers of today—quite a historical journey! As wonderful as these base metals, minerals, and organic polymers have been to our advancement, they also have always been of limited plasticity and, therefore, of limited use. It wasn't until we learned how to harness the unique material found in the ground as a superpolymer when our definition of a material would change, as it is today. It was a hell of a packaged deal: an energy source so efficient and powerful it led to the manufacturing process of *heat, beat,* and *treat* and, as a result, to the divergence of society from using natural materials to synthetic materials, where things would never be the same.

At the onset of the 20th century, chemists were beginning to understand that natural resins and fibers were polymeric, meaning they were formed by subunits of monomers. Having gained this greater knowledge of chemistry and the powerful source of fossil fuel, material scientists began to feel God-like, able to create synthetic polymers more flexible, useful, and inexpensive than the natural ones. In 1907, incented by money rather than fame, Leo Baekeland sought a synthetic replacement for shellac, a natural resin found in trees from a secretion from lac bugs and used primarily for a wood finish. Instead, what he accidentally stumbled upon in the combination of formaldehyde and phenol, when placed under high temperature and pressure, was something infinitely more useful: the first synthetic plastic that was heat resistant and nonconductive, making it very useful for insulation and casings for new devices such as radios and telephones. What was magical about Bakelite was its many purposes; it could be used for applications in jewelry, toys, electrical and insulation products, floors and ceiling tiles, and others.[3] No material ever made, natural or synthetic, was so multidimensional, useful, scalable, and affordable, leading to a material revolution through polymer science. What was not known about these synthetic polymers, however, would be discovered later on. Bakelite—made from formaldehyde, asbestos, and phenol, a waste product from coal—can cause cancer and mesothelioma from the asbestos that leeched from it and amyotrophic lateral sclerosis (ALS) from the formaldehyde.[4] Yet this was the first synthetic polymer known to us; the plastic that started it all. In 1924, *Time* magazine predicted that Bakelite, "the material of a thousand uses," would make up everything we touch, see, and use, and no doubt that prediction came true in all the different forms of plastic that were to follow.

From this point, polymer chemistry and its applications took off. Polyvinyl chloride, or PVC, was developed in 1927 by B.F. Goodrich as well as synthetic rubber to displace rubber tree plants in 1931. Polystyrene (PS) was developed in 1941 by Dow Chemical, notably for the use of Styrofoam. DuPont's Wallace Carothers invented Nylon 66 (Nylon) in 1935 as a replacement for silk, and in 1941, low-density polyethylene (LDPE) was invented. The DuPont Company, a company founded on the production of gunpowder, had an interest, and in 1927, took aim at a quirky yet brilliant Harvard academic named Wallace Carothers, who helped lead the creation of neoprene in 1928. Polyesters, polyamides, and nylons in the 1930s were discoveries in the world of the unknown that led to creations once thought beyond the realm of possibilities. Quickly, synthetic polymers became serious business, as *Time* noted, "Plastics have been turned into new uses and the adaptability of plastics demonstrated all over again."[5] During World War II, plastic production increased by 300 percent as a matter of meeting the needs for the war effort and to supplement materials at home.[6] During this era, plastic really came into its own, both in breadth and depth. America would transition from a wartime economy, creating materials necessary for victory, to a peacetime one by these materials transitioned from military to civilian use. As a result, plastic was the key material for not only winning the war but also securing the peace.

Plastic Alchemy 2.0

Plastic was probably as critically important to winning the war for the Allies as it was for securing the peace afterwards. During the war, metals of steel, copper, aluminum, and zinc became precious commodities too few, and in order to supply the war, materials such as cellulosic, acrylic, nylon, phenol, and polyethylene were manufactured at large scale for the war effort to succeed.[7] In a time of great need, there is often great innovation; America solved its material crisis back then, and it can solve it today as well. After the war, the question became how America could win the peace, how wartime material such as inflexible thermosets could become useful as thermoplastics, transitioning the first generation of plastic (Bakelite) into the next era, the one we are more familiar with, which consists primarily of polyethylene and polystyrene. A thermoset plastic like Bakelite is strengthened by a strong chemical bond in the production process, but it doesn't possess as much versatility for new product design as a result and is difficult, if not impossible, to recycle and reuse. The benefits of a thermoset plastic is once it is set, it cannot be remolded, making it sturdier under high temperature. It is generally less costly than thermoplastics and has less potential for leeching, a process where chemicals are unintentionally released from the material. Due to its limited capabilities, thermosets are used in lower demand than thermoplastics and mainly relegated to such traditional uses as wire insulation, epoxies, rubbers, and polyesters. In contrast, thermoplastics are believed to be more flexible and useful, are more recyclable, but are also more likely to leech into the environment when not reused. The thermoplastics that are most familiar are polyethylene (PE), polypropylene (PP), polystyrene (PS), polyvinyl chlorine (PVC), and polyethylene terephthalate (PET). These thermoplastics are more recyclable than thermosets, nothing to brag about since thermosets are not recycled at all! Due to their looser bonds, thermoplastics are alleged to be recyclable, but they become a greater problem in nature when they are not. Most important to the plastics industry, thermoplastics have been the driver of growth since World War II due to significant innovations in both variation of designs and capability to scale at low cost what is produced. Yet it would be fair to say the growth of production of thermoplastics has vastly outpaced our knowledge of how to reuse these materials or their afterlife if not reused, a bit of unknown alchemy the wizards never expected or even considered.

How plastic became the savior of the 20th century consumer society after World War II, however, is no mystery. Start with a gigantic war machine no longer in need and add a citizen base that has been subject to sacrifice and rations first through the Great Depression and then through World War II, and the solution is abundantly obvious: build a great big consumer economy to replace the giant war machine, enabling a multiplier effect between supply and demand, driven by supply chain innovation. Looking back over the last century, there is no doubt plastic production and consumption have become a key cog in a consumption-based economy. In 1950, there was a little more than 1.5 million tons of plastic produced worldwide; 10 years later, it would quadruple to 6 million tons, and it would nearly quadruple for the next two decades (the 1970s and 1980s), and almost double the next three decades after that (1990–present).[8] Plastic became the modern day philosopher's stone, the magic able to turn depression and war into leisure and convenience, first through military equipment, and then through a bunch of stuff for pent-up consumers who were ready to break free. In the end, however, it may be a Faustian contract, a deal with the Devil, when we learn of its consequences, as is being addressed in this book. Once our obsession with plastic started, it could not be stopped nor contained; a consumer broke free from nature, obtaining the gadgets and conveniences never imagined that led us to be hooked on the stuff. Today, we are learning that we may have to give the Devil his due one day; we will have to settle up. It's just a matter of when this will happen.

Sorting Out the Acronyms

When I started this research on plastic, I began to poll people I knew, including those in the sciences, to ask them about their knowledge regarding the myriad types of plastic resins, recipes, and products that are produced, among us everywhere we go, and then left to rot at staggering degrees of volume. Not just the average consumer but even the consumer-products companies that use these plastic resins for their products and packaging materials know very little other than the requirements for their businesses. Through a supply chain of specialization, each agent in the system knows only what is relevant to its role, and this often means they are unfamiliar with the formulas and recipes of the resins they are using and, more importantly, any repercussions through its use by a consumer, or even those who choose not to use, but are exposed in some manner. In reality, I don't think there is anyone who really has a sufficient understanding of this system (as opposed to the material) I'm calling plastic, a system that is almost indescribable. So even though I place myself in that category with everyone else, I will continue this chapter with a primer on plastic, so do not shoot the messenger for attempting to explain it!

There are the three types of plastic that comprise almost 70 percent of the total market: polyethylene (36 percent), polypropylene (21 percent), and polyvinyl chloride (12 percent).[9] Of the remaining plastics, there are a few others that are well known, such as polyethylene terephthalate, which makes up your disposable plastic bottles, that is 7 percent of the total world market and polystyrene, which makes up that unique thing called Styrofoam and is used for soda takeout cups and disposable coolers, that is also 7 percent. Lastly, there is a bucket of plastic types classified as "miscellaneous" because they do not fit into any one category and could be practically anything. This high-level categorization of plastic refers to its "base polymer," a rule of thumb more than anything regarding how these resins are designed as well as their impact to us and our environment throughout the supply chain and post-use.

Plastic #1, identified by the chasing arrows on your plastic water bottle, is polyethylene terephthalate, or PET, a thermoplastic polymer used for clothing (60 percent), plastic bottles (30 percent), and other products. What makes this plastic useful in structural applications is its glass-like appearance, strength, ability to stretch, light weight, and effective barrier properties for both water and gases. This plastic is formed using a polymerization method called condensation that makes it easier to recycle than some of the other types of plastic. PET plastic is recycled at a higher rate than any other plastic, but the bad news is that rate is still only at 20–30 percent, with much of PET being downcycled in its reuse, a nice way of saying it is used for lessor purposes, such as fake-wood decking rather than bottle-to-bottle reuse. Due to its low recycling rate, a significant amount of virgin PET resin is needed to constantly be resupplied into the process to make new bottles, therefore further exacerbating the problem of a linear versus closed-loop system. In Scandinavia, where I conducted my doctoral research, PET recycling rates are higher, but this is due to government mandates in place requiring recycling to eliminate the need for landfilling rather than a market-based justification for these used containers in a closed-loop system based on viable economics. In reality, what most people do not know is bottle-to-bottle reuse rates in Sweden is not much different than they are in the United States, where collection rates are only 20 percent versus the 80–90 percent in Sweden. The difference between these rates is Sweden does not landfill its excess material like we do in the United States, and the *reuse* is loosely defined in a sophisticated clean waste to energy scheme where PET waste is incinerated for electricity, not what most people think of when they hear the term "zero waste." These expensive modes of eliminating an abundance of plastic bottle waste happens in a few nations but not in most; as a result, our world is closing in on 500 billion plastic bottles used annually, with a large majority of them never even attempted at being reused after its fifteen minutes of fame.

Despite these statistics, and there are many more, it is customary in most cultures to believe that "all plastic bottles can be and should be recycled" as a matter of fact. Yet, the science and economics of the reality tells a much different story. In my household, it is my job to take our used plastic bottles out to the garage to the big green commingled recycling bin despite my lectures regarding its futility of what really happens as nothing more than Kabuki theater, not to mention the added cost to our costs and those of the waste management company. Yet to my wife and daughters, recycling is a civic obligation that must be done, at the very least because it makes them feel better, which is what they finally tell me so I'll stop talking about it. Who cares if much of what is recycled is eventually thrown into a landfill? The truth is recycling could work if the overall supply chain system existed in a manner where these plastic bottles were more economic to reuse; because the product and its corresponding supply chain system was not designed this way at the onset, we are left with a culture of shaming and greenwashing that makes us feel guilty regarding our consumer behavior and becomes a distraction from potential solutions that must address the root cause problems that I will discuss in this book. For those who challenge this point of view, and there have been at least a few people who have labeled me as cynical for such a perspective, start at today's plastic commodity market and see how the fetch price of recycled PET, or rPET, is much lower than that of virgin PET due to a lack of demand for the former in comparison to the latter; if rPET is more sustainable and efficient, why is this the case? In contrast, the aluminum market is the exact opposite; recycled aluminum is more efficient and in demand than virgin aluminum due to greater sustainable and lower conversion costs. Furthermore, while recycled aluminum has greater ease of processing and similar structural integrity as its virgin product, rPET does not, meaning it's less efficient to reuse in the supply chain system. The conventional method of recycling these bottles is a process called "mechanical recycling," the cleaning and grinding of the plastic in order for it to be reused. Mechanical recycling is the lowest-cost method for rPET production, but it is not very efficient; the effective material yield from this process is rather low at 25–50 percent, higher in places like Sweden that mandate a costly collection and processing of recyclables that leads to a pristine condition of the used beverage container. In other markets, such as the United States and developing nations that do not mandate such care of recyclables, a lower yield rate is achieved, requiring a higher percentage of virgin PET added to the mix, meaning it's often not worth the trouble to include recycled content other than for marketing purposes.

Recycling enthusiasts, particularly from Europe, can be especially defensive toward any challenge to the legitimacy of PET mechanical recycling given how invested they are in this method, both monetarily and culturally. Of course, it is a better alternative to doing nothing, but it is questionable whether it is the most effective approach from an overall supply chain perspective given the upfront costs associated with handling this waste in low market demand in such a pristine manner. The high costs of mandated recycling, the low yields of mechanical recycling, and the corresponding high percentage of virgin PET required for new plastic bottle production should be the impetus for more transformative solutions that will be framed in the final chapters of this book. Not only is mechanical recycling inefficient, it is bad science; it is a process to reuse the polymers already intact, and as such, it limits the structural integrity of the material thus requiring a higher percentage of virgin PET to strengthen and improve the polymer. A better alternative, and the solution I will discuss in chapter 8, is that of chemical recycling, the actual de-polymerization of the PET plastic back into its original monomers through a catalysis process. Chemical recycling, by definition, leads to much higher material yields but at a higher conversion cost than mechanical recycling, which is the reason why it is not widely used. However, looking at this problem differently through the use of supply chain innovations to create a new system approach that I will also discuss in chapter 8, chemical recycling can be cost effective in comparison to mechanical recycling, leading to a potential transformation to solve this problem rather than to simply mitigate it, as is the framework for today's approach. In the solutions chapters of the book, I will discuss how today's approach of mechanical recycling is just mitigating waste and limiting recycling rates versus a chemical recycling approach within a transformed supply chain system can offer true environmental sustainability and a near closed-loop system of reuse, like nature. For now, you should know the limited design of PET (not designed to be recycled), the inefficient conversion process of mechanical recycling, and the lack of an end-to-end supply chain system for plastic bottles are the primary reasons for why recycling rates are so low, not because the consumer is too lazy to place a plastic bottle into a recycling bin.

For the most part, PET plastic is considered to be safe for use with food and drink; however, there is some research that raises health concerns that should be studied in greater depth. In comparison to other plastic resins, the base monomers in the design of ethylene glycol (MEG) and terephthalic acid (TPA) that make up PET appear to be manufactured without a significant amount of chemical additives, but concerns do exist. Antimony, a metalloid additive with known toxic qualities, is an additive and has been found to leech into containers with the potential of contaminating the liquid, particularly at higher temperature levels. Environmental Protection Agency (EPA) tests have found the antimony levels that seeped into the liquid are well below the acceptable standards, but whether these tests were conducted addressing all the possibilities in the supply chain is unclear, such as under higher levels of temperature during distribution, storage, use, and perhaps the long-term effect of smaller doses on the body. This is a critical question given today's bottled water supply chain where storage conditions possible for bottled water and other beverages that are shipped for hundreds or even thousands of miles are without any temperature control during the summer.[10] In other studies, high temperatures of around 140 degrees Fahrenheit for 176 days has led to unhealthy levels of antimony above the approved EPA levels, and while these conditions are unlikely even in the summertime in our garages, where is the line drawn for these conditions or if a consumer microwaves a PET container without understanding the inherent danger of doing so?[11]

These research questions are largely unresolved: Should the consumer be concerned or possess a reasonable doubt related to its safety in all uses? Under what conditions is a PET plastic beverage container safe when the liquid-bottle combination is under different circumstances throughout the entire supply chain system? According to the Center for Disease Control (CDC), exposure to antimony can lead to health issues such as diarrhea, vomiting, and stomach ulcers, but again the question of safety is related to what exposure at various doses and levels. It is also recommended plastic bottle manufacturers wash bottles to rid them of contaminants like antimony and bromines, but there are no requirements to do so.[12] In a 2013 study performed by German researchers, nearly 25,000 chemicals were found in a single bottle of water, many of which are concerning for various health reasons.[13] From one perspective, this is of little concern, especially if the greater good is a bottle of water used as a portable source of hydration, an important necessity for the human body. But from another perspective, do we really understand these alleged research concerns, especially when so many consumers believed often incorrectly that it is a healthier choice to tap water (even filtered)? From my perspective, it isn't a matter of which side is right or wrong as much as it is to better understand this product that is so important in today's consumer marketplace.

Perhaps, the use of the PET plastic bottle as a portable delivery device for clean water has become a catch-22 in modern society that we do not understand it to be: it can be a lifesaver to communities both in the developing and developed world that are either lacking a stable and sufficient municipal water system or in the midst of a failing system that is becoming more of a prominent problem today in the developed world, such as what is happening in Flint, Michigan. Yet on the other side, it can be a largely unregulated and under-researched product, without sufficient evidence to guarantee safety. In times of urgent need, bottled water becomes an emergency crutch to these undeveloped, insufficient, or failing water systems, but can trade the problem of an inadequate water supply for a dependency on single-use plastic as an out-of-control waste management problem that the municipalities may be equally unable to address. In the United States, we have seen how this problem can take root, such as bottled water in Flint, Michigan, and Puerto Rico to solve a water crisis while piling up unrecyclable bottles, a dilemma of our modern-day supply chain system; clean water unable to be distributed to citizens through public infrastructure is humped, liter by liter, through an excessive and unending stream of PET plastic waste that is recycled at a rate of 20 percent or lower.

The use of the PET bottle as a delivery mechanism for safe water in the 21st century will only grow as a phenomenon around the world, and as this happens, there are questions that must be addressed. For one, if bottled water use around the world grows largely in developing nations with less capability to recycle, how will a worldwide recycling rate of, let's say, 20 percent and a worldwide annual consumption volume of 500 billion impact our environment and bodies at such growing scale? At these levels, spread across the world, how can researchers understand and analyze the integrative effect, causal, correlational, contributing, or otherwise, to our health and the health of the planet? It seems to be an impossible task, especially given the lack of precise testing procedures and tools to detect the influence of micro- and nano-plastics to our molecular and genetic make-up, especially as we are more surrounded by its products in all of its forms. Without sufficient evidence and controls in place to ensure safety at this measurement level, there should be a measure of concern regarding exposure to antimony, bromines, and even acetaldehyde, a known human carcinogen. Without prejudgment and an appreciation of the good that plastic brings to us, our goal should not be to overreact nor to deny the scale of its existence of these resins and formulas as an acceptance risk without a better understanding. And particularly of importance is the need to design a new supply chain system of better safety and reuse to resolve the problems that we don't, or perhaps can never, understand as full as we would like to as consumers.

Polyethylene (PE) is the largest plastic used in the world, comprising of almost one-third of all plastics produced and sold due to its toughness, flexibility, and ease of processing. It is a waxy material of lower strength, yet has hardness and rigidity that are useful in a variety of factors related to its density level. Legend has it John D. Rockefeller once looked outside the window of this office and saw ethylene being burned off through flames shooting up at the oil refinery and told his people to find a use for it, and they did through the development of polyethylene. It is said to be stiffer than steel and softer than wax with a relatively simple compound of repeated methylene chains. When these chains are catalyzed at a moderate level of pressure, it forms high-density polyethylene (HDPE or #2 plastic), and at higher pressures, low-density polyethylene (LDPE or #4 plastic). HDPE makes up 15 percent of the global market, and it is a thermoplastic like PET but an addition polymer, meaning its polymerization process is different, and it is not as easy to recycle through chemical means. The polymer chains of HDPE are packed tighter together than LDPE, giving it a higher tensile strength that is useful in such consumer products as grocery bags, milk jugs, and detergent containers as well as water drainpipes in construction. In comparison, LDPE (#4 plastic) makes up 17 percent of the overall global market and is used in food containers, sandwich bags, computer components, and now as plastic wraps as a replacement to polyvinyl chloride that contains phthalates. Alongside PET, HDPE is the highest recycled plastic, having a similar 20–30 percent recycling rate, with the two accounting for 96 percent of all plastics recycled in the United States.[14] That these two groups of plastic polymers make up approximately a quarter of the overall market but 96 percent of materials recycled is a shocking statistic. Similar to PET recycling, HDPE is mechanically recycled (i.e., polymers ground rather than depolymerized), limiting its economic opportunities toward a viable closed-loop system, which means this method will never solve our plastic problem. Unlike PET, a condensation polymer that is relatively easier to reconstruct to its original monomers, polyethylene is an addition polymer that cannot yet be cost effectively recycled chemically, an issue I will address in the solutions chapters of the book. Suffice to say at this point that polyethylene has the potential to be a highly recyclable plastic despite HDPE being currently around 25–30 percent and LDPE at 0.8 percent; a disruptive innovation in the marketplace is the only salvation for the future. Given the large variation in the different uses of plastics within even one category, and the lack of knowledge and public exposure required to address the problem, the status quo is incapable of changing this paradigm.

Plastic #5 is polypropylene (PP), a plastic that makes up 23 percent of the world market and is a thermoplastic addition polymer. It is very similar to polyethylene in its composition, but it has a monomer of propylene versus ethylene, and it can be clear in design where an ethylene can only be translucent. Polypropylene has a higher melting point than polyethylene, but PP does not stand up in cold weather as well as PE. As well, polypropylene is less stretchy than PE, better resistant to chemicals and solvents, but not as stable. PP is categorized as stretchy (but not as much as PE) and moisture proof, making it useful for bottle caps, storage bins, to-go containers for food and drink, disposable diapers, shrink-wrap, and much of the plastic in a car. Its recycling rate is very low because it is often not used as a resin by itself but is mixed with other plastics, making it difficult and costly to recycle.[15] This is a critical point to understand regarding the design and recycling of plastics: the ability of polymer scientists to create novel combinations to serve consumer need has far exceeded an end-to-end supply chain perspective of how to handle all this stuff post-use. As a result, today's plastic recycling capabilities is lagging far behind keeping pace from a scale and technical standpoint, which is a significant driver of the Peak Plastic problem mentioned in this book.

Similar to PET, polyethylene (PE) and polypropylene (PP) are considered low-hazard plastics, but this may only be in comparison to the more dangerous resins I will discuss later in this chapter. For example, low-density polyethylene (LDPE) is being used more often as a plastic food wrap than is polyvinyl chloride (PVC) that was used before, but it still may contain a chemical called diethylhexyl adipate (DEHA), a suspected endocrine disruptor that can cause issues if it leeches into your food during the microwaving process. Because manufacturers are not required by law to list the chemical ingredients on the label of its products, such as a plastic food wrap, there is no way of knowing what is included or if it is safe and in what use. How much do we really know about these designs, and why are there not more concerns raised by the general public? It seems to be a problem too complex to understand and resolve: resin formulas may be categorized as a HPDE, LDPE, or PP, but these provide little information other than a very broad classification. To my understanding, there is no access to specific ingredient disclosures despite studies conducted that have found plastic products, including HDPE, to release estrogenic activity in some causal or correlative relationship when microwaved, exposed to sunlight, or during other basic chemical processes, such as boiling water.[15]

Other studies have found polypropylene to release measurable levels of butylated hydroxytoluene (BHT), but a definition of measurable levels under what conditions is unclear, as is which specific PP formula/manufacturer was used in the study.[16] Furthermore, with the tools being used, it seems to be very difficult to isolate how a specific chemical within a specific plastic resin induces estrogenic activity, and whether that specific chemical was a causal, correlative, contributing, or mere coincidental factor. Maybe in the future with an implementation of technology to better interpret molecular level information or genetics, a greater understanding could be possible at a submicroscopic level. Scientists and researchers seem to hold more questions than answers regarding the overall health risks, not because there is not cause for concern but rather due to the difficulty of isolating a specific cause and effect. One study found nonylphenol (NP) leeching into milk from an

HDPE container similar to levels found in water from a plastic bottle.[17] Specific studies under specific conditions lead to questions without formative conclusions at this point, leading to a conundrum of what to do next. This means while no conclusive evidence suggests a danger in using these safer plastics of PET, HDPE, LDPE, and PP in any manner, it is a general seal of approval due to a lack of harm confirmed in testing. The impact of conditions such as heat, longitudinal impact, and an inability to validate what chemicals may linger in the end product from the manufacturing process is not evident. Without much, if any, public exposure to the potential problems, there are no proposals on the table to conduct greater testing or information disclosure of formulas to consumers, rendering them as safe with no conditions.

Plastic #3 is polyvinyl chloride (PVC), a thermoplastic addition polymer that makes up 16 percent of the total world market. It is very useful due to its unique combination of being lightweight, strong, cheap, and of low reactivity. As a result, its most frequent use is in the building and construction sector, where 69 percent of all PVC is used, two thirds of which is in piping; other uses include house siding, wire insulation (a replacement for Bakelite), fire retardant clothing, shower curtains, medical devices, and pleather, the fake leather in your car seating. It is often defined as being shape-shifty, meaning it is rigid but flexible due to chlorine atoms at the end of its polyethylene chains that make it more versatile. The base polymer itself is brittle, and to make it shape-shifty, some questionable additives are included in the process, which are unidentified when you purchase the product. Some products made from PVC, such as food shrink-wrap, have been replaced with other plastic polymers, but it is still being used in many ways where it can come in human contact or consumption. Because there are many variations and questionable additives that are included in its ingredients, this plastic is logistically impossible and economically impractical to recycle. As a result, it holds the dubious reputation of being uniquely qualified in so many different types of applications given its lightweight and high-strength nature but suspicious in consumer use and post-use given a lack of recyclability and high toxicity concerns. If these materials are potentially too dangerous for use in a microwave with food, is it safe to leech in a landfill or waterway, or as a seat in your car? The answer to this question isn't clear, but there are sufficient concerns to raise the question. PVC is sometimes called the "poison plastic" because many of its additive plasticizers have been associated with cancer, birth defects, genetic changes, bronchitis, ulcers, skin diseases, deafness, vision failure, indigestion, and liver dysfunction.[18] Because of the large use of chlorines in its production, PVC manufacturing releases dioxins, one of the worst carcinogens in use. After 27 years of review, the EPA released the first part of an anticipated study that linked the use of dioxins from PVC manufacturing facilities to non-cancer health effects such as developmental and reproductive problems, damage to the immune system, and hormone disruption.[19] By definition of its base polymer being rigid and needing to be modified to become unique, PVC can include toxins like lead and cadmium in its ingredient list of plasticizers and stabilizers, harkening back to Dr. Frankenstein's approach to manmade creations that have unintended consequences. Unfortunately, while such plasticizers as phthalates soften the plastic for form and function, these additives do not completely bind with the polymer, making them highly susceptible to leeching, inadvertent human digestion, and groundwater contamination among some of the possibilities. PVC is perhaps the poster child for why the design process for plastic has missed a link that hasn't accounted for the entire supply chain life cycle, including our ecosystems and bodies.

Even if the regulatory bodies called for a complete ban and suspended PVC use immediately, I wonder to what extent we would be capable of controlling its impact since PVC is used in so many different products and incalculable recipes that it would be seemingly impossible to measure, not including the tons of this stuff rotting in a landfill or already escaped into the ecosystem. As an example, while there are some regulations in the United States and the European Union to manage some concerns, such as the use of lead as an additive, not all have been controlled for, and the same regulations do not exist in manufacturing nations (e.g., China) where so much of the world's products are produced. There are a significant number of studies that have found a correlation between PVC and health risks without determining a clear cause and effect, leading to a quandary of how to interpret what action to take or not based on such findings. For example, a 2008 Swedish study found that plasticized PVC leeches toxic chemicals into water, but the study did not assess the impact of these chemicals on humans, leading to the question of whether it is dangerous or not.[20] Other studies have found the phthalates from PVC to have adverse effects on airways and immunologic systems, but without any substantive conclusions regarding its impact on humans. Even your shower curtain made from PVC has been found to release 108 volatile organic chemicals into the air that can cause serious health risks to humans.[21] The design, manufacturing, use, and post-use of PVC is a mystery to nearly all of us and has the potential of being improperly managed as a result. The plastic supply chain innovation engine over the past 70 years of producing better products cheaper continues to chug along without an equal balance from an environmental, health, and safety standpoint. Today, these supply chain systems are left to their own devices and considered as separate from our public policy where these questions are sometimes being asked. As the scale and complexity of these supply chain systems continue to grow, including a lack of standardization across nations regarding regulations and control, the problems, intentional and unmeasured, will grow with us, leading to this Peak Plastic predicament discussed in this book.

The next plastic is polystyrene (PS or #6), which makes up 7 percent of the world market. The base model is a rigid plastic, like disposable food utensils and silverware, casings for a variety of appliances, insulation, bike helmets food service containers, and portable coolers. The other main category material is often known as its form when it is puffed up with air (95 percent air to 5 percent material) to make Styrofoam, a brand-name material from Dow Chemical. There are 25 billion Styrofoam cups thrown away every year just in the United States, and none of them are recycled.[22] While there is some recycling of standard polystyrene plastic, there is no collection for its expanded polystyrene materials (i.e., foams) mainly due to a lack of demand, and bad economics, if not largely due to its poor design in consideration of a closed-loop system. As a result, an increasing number of communities in the United States and other nations are banning the expanded foam polystyrene Styrofoam given the inability to recycle the material and the corresponding health concerns. Similar to LDPE, PP, and PVC, its large variety of uses makes it impractical to recycle given today's methods and financial models.

Similar to the story of other plastics, it is inconclusive, at best, whether polystyrene is safe for use; in 2013, the Plastics Foodservice Packaging Group provided updated styrene migration data to the FDA and found the current exposures to be extremely low, at 6.6 micrograms a day, more than 10,000 times lower than the safety limit imposed by its regulations.[23] However, polystyrene, made from benzene and styrene, two known human carcinogens when handled improperly, was not used in the test of this material (styrene) by the packaging group noted above. Styrene has been linked to cancer, but the EPA has yet to provide it with a formal carcinogen classification, even though it admits there are links between the material and the disease, among other potential concerns.[24] Some studies have shown styrene to be a neurotoxin over the long-term, with studies reporting harmful effects on animals.[25] Studies have also shown temperature can play a major role in the leeching of styrene, and polystyrene should be avoided in food packaging; but of course, there are many factors to consider, including what temperature is required to instigate leeching, how oily the food is at a given temperature level, and whether the amount of leeched material is sufficiently dangerous to the user.[26] It once again raises the question as to how to study these materials and their effect on us and the environment, and what is the proper action to undertake if conclusions are difficult to discern: that is, whether to assume the worst- or best-case scenario from a public policy standpoint.

All the other permutations are called plastic #7, a catchall category of the remaining plastic universe. It is a meaningless category other than to state none of these are plastics #1–#6 other than perhaps being combinations of them in some form or fashion. Of course, none of these formulas are attempted to be recycled given the mysterious nature of what is included in them. In comparison, PET and HDPE plastics are the highest recycled polymers because they are typically the most homogenous. For the most part, a PET plastic bottle is a standard formulation, and as a result, a bottle from one manufacturer can be combined with another during the recycling process with confidence; this of course is not the case for other plastics, and definitely not any product with a #7 on its label. One notable ingredient in #7 plastic has been bisphenol A (BPA), a bad hombre suspected to cause problems such as hormone disruption in humans. Chances are you have it in your body, as 95 percent of all humans do, but it is not conclusive of the impact it may have to your health. In just one week in using polycarbonate plastic, such as reusable water bottles (advertised as helping rid the earth of disposable plastic bottles, ironically), BPA can increase urinary concentration of the chemical by 66 percent.[27]

Polycarbonate (PC), the most well known of these other plastics, is a tough and transparent material, making it very useful in lab equipment, eyeglass lenses, lining in can foods, and, in the past, baby and water bottles. Because it is so tough—stronger than even PVC—but also easy to mold, it has been useful in a variety of applications. It is a thermoset plastic, however, which means it can only be heated once, and as a result is difficult to recycle. The concern with PC is the release of BPA due to hydrolysis, a chemical reaction of water. The most commonly formed general ingredients of polycarbonate are BPA with phosgene which is an effort to decrease the effects of BPA without a great understanding as to whether this will make it safer or not. The leeching of BPA is subject to human health concern but not well understood as to how it breaks down and impacts normal hormone function in the human body. Studies have shown very low doses of it can cause quite a few health risks, such as cancers, impaired immune functions, early onset of puberty, obesity, diabetes, and others.[28]

This miscellaneous grouping of plastics is not well known by name, but certainly by use, such as the polyurethane in cushioning for furniture, linings, and condoms; acrylonitrile-butadiene-styrene most notable in Legos but also found as plastic musical instruments and phone casings; phenolics used in cutlery handles and electrical insulation; nylon used mainly in clothing; and acrylic used for airplane windows and car taillights. The tens of billions of Lego bricks manufactured and sold every year are not recycled, as the billions of used condoms, and there is also the bigger question regarding the safety of these materials, such as a Lego in a toddler's mouth or others when released into the environment. Formaldehyde, our old friend from Bakelite, is still used today in industrial and medical applications, such as a disinfectant to kill many types of bacteria, and has been known for decades to be a carcinogen. Polyurethane used for cushions, mattresses, and pillows can release something called "toluene" that can lead to lung problems, as can acrylic used in blankets, contact lenses, diapers, sanitary napkins, and so many other products.[29] And the list goes on and on regarding products with traces or significant amounts of plastics we do not understand or are not classified at all in the listing of the product for consumer protection. For some of these products, its trace chemicals are tested regarding its impact on the human body, while some are not; certainly the cumulative impact of them all together, including that of the first six categories of plastic, is not tested, as that would be an impossibility. What this means is that while it is not possible to understand the full impact of all of the plastics in use regarding their impact to us and the environment, it doesn't necessarily mean there is a clean bill of health as a result, and it would be foolish to assume so for matters of convenience.

What Lurks Beneath the Polymer

There is really no such thing as *plastic* other than these broad categorizations to help our understanding. Even the six categorizations, with the seventh as a catchall, is limiting; for each of these categories, there are subsections (e.g., hardened polystyrene, known as PS, versus foamed polystyrene, known as extended polystyrene or EPS). Even how safe these materials are is only a generalization, with the first four resins of PET, HDPE, LDPE, and PP considered safe in comparison to the greater concerns from PVC, PS, and PC or because there is insufficient evidence to point in one direction or the other. Considering the safety of a material given some of these additives is very difficult to assess because we must take into account such factors as the scale of its use and dose and in what conditions it is being used, (e.g., a case of plastic water bottles sitting in a car trunk in August in Phoenix, Arizona, or something plastic placed into a microwave). In other cases, there is a need for generational studies, such as with PVC, and the dose and extent of our use is so new, evolving, and scaling as to be difficult to assess. Finally, how can we assess the impact of all of these synthetic substances to our bodies in totality? What seems to be clear though is as our use of these synthetics grows in scope and scale across the planet, there should be a greater, not lesser, concern on how to manage it, which is something that is lacking across these global supply chain systems. So it's not a point of can we make broad statements as to whether a plastic is safe or not but whether we possess a sufficient understanding to make a judgment—and the answer to this is clearly no. Our stance should be that our knowledge should grow as our use and complexity does, and yet the opposite is clearly happening today.

As a general rule of thumb, the base polymer compounds are made primarily from carbon atoms and include oxygen, nitrogen, and sulfur, which are artificial replications of natural polymers and are generally deemed as safe in the environment and our bodies. Some combinations and monomers get suspect, such as styrene as a potential cancer-causing agent and PVC with its chloride monomer causing concern. The biggest concern with base polymers seems to be the additives to the base that are smaller molecules required to achieve the desired design but potentially toxic in certain circumstances or doses.[30] For the most part, even those trained in chemistry will refer to these resins by their base polymers when describing what makes each plastic unique while at the same time not paying sufficient attention to any additives, which may not be identified in the design but are of greatest concern. In reality, it is often these final additives that have made plastic "plastic" through creating these infinite degrees of specification to achieve form and function at the lowest cost. Given this challenge, it seems logical for an improvement in our public understanding of what exactly these materials are made from in their bills of material and the manufacturing process. One study found that 98 percent of all plastics off-gas or leech some of the monomer, leading to an impact on us and our environment that complicates any understanding.[16] Once a product is used, a substance can be released, and once these products proliferate through widespread consumer use, it may be impossible to separate what's good from bad when tens of millions of tons are used annually and compounded over decades across every nook and cranny of the Earth.

What lurks beneath these conceptually clean polymer definitions are the additives used as stabilizers: antistatic agents, flame retardants, plasticizers, lubricants, slip agents, curing agents, foaming agents, biocides, colorants, fillers, and reinforcements, among others.[30] Some of these agents are included in small quantities compared to the total material by weight, such as stabilizers that can be as low as 0.05–3 percent of a final product, but flame retardants can be 12–18 percent by weight of a product, and plasticizers can be 10–70 percent, the highest amount in PVC. According to one study, 13 out of 16 homes in Northern California were tested to have flame retardants in the dust we breathe, causing concerns.[31] Because so many additives are embedded in ordinary products and leech into dust in the air we breathe, you probably have no idea what might be lurking in your home, but one chemical you may have heard of is phthalate, sometimes known as the "everywhere chemical." This chemical is used primarily as a plasticizer, most often publicized for its use in food and beverage containers (yet is included in practically all plastic products), and can also be attributed to the new-car smell so many seek when buying a vehicle. A billion pounds of phthalate is produced and used in a year, which is probably the reason why 95 percent of us have it at detectable levels in our bodies.[32] No doubt how we are being exposed to it today is various, misunderstood, and not advertised—including traces found in tap water and in the form of pesticides on our fruits and vegetables—but that doesn't mean we are in danger or that we are safe.

Despite phthalates being the most infamous of these additives, there is no clear consensus as to whether it is safe; according to a recent 2018 study released by the National Toxicology Program in the United States, BPA is safe and not making us sick.[33] This finding was consistent to a 2014 study that concluded there is no clear effect of phthalates on human health, which has set off a debate within the research community regarding the validity of conflicting studies. Much of the debate surrounds such significant variables as temperature as to why some studies and sources raise concerns while others do not. In both the United States and European Union, and elsewhere, phthalates have been banned in some uses, but its use is so widespread in so many plastics it leads to questions as to why it is safe in some products, or under some conditions, and not others. As a result of being virtually everywhere, it leads to more questions than answers, and it is probably close to being impossible to study in today's society where it is so widespread. One notable phthalate frequently brought up as a possible carcinogen and toxin is di(2-ethylhexyl) phthalate, or DEHP, which has been banned in Europe and the United States for use in baby toys but not from many other functions, such as IV bags where babies can get an even more direct exposure. We once again face this question as to whether it is safe or unsafe in some circumstances and not others, to some sub-segments of the population (babies and elderly) versus others, and whether tiny impacts have a cumulative effect on top of other synthetic materials. It also leads to a question of how today's supply chain system is responsible for producing so many goods so efficiently yet posing dangers in such an invisible sort of manner, and how any of this is tracked in a global supply chain system.

BPA has gotten a lot of attention regarding its use in reusable water bottles but not as much in other products such as storage containers, can linings, epoxies, resins, and coatings, and it has been in use since the late 1950s. It seems like this is an "ask and answer" problem, meaning other products, such as the lining for a food container, have not been raised to the same level as a water bottle, which is not necessarily an affirmation those products are any safer. Studies have shown a large percentage of us have BPA in our bodies; between 2007 and 2009, 91 percent of Canadians tested had BPA in their urine.[34] Currently, this chemical family is considered as safe for all uses by the FDA despite these concerns from a number of studies that suggest it to be a disruption to hormones in animals among other dangers. A campaign is underway to remove these chemicals from some products, with different adoption policies in the United States versus the European Union and certainly more lax consideration in Asia, without any rhyme or reason in how products and chemicals flow in our modern supply chain system and into our natural ecosystems. Given the global nature of our supply chain systems, how can consumers from one nation be assured higher standards are enacted to products produced with less stringent standards? The answer to this question is not very clear.

As these chemicals are found in our natural water systems like rivers, lakes, oceans, and even reservoirs, they are suspected of altering the mating habits of the fish we eat and have an impact to other species as well. Given the significant role of Asia as a seafood exporter to the West, specific national or regional regulations become unclear, less effective, or even nullified within a global supply chain system. An inability to track the environmental, health, and safety connections within our modern supply chains related to the impact to and from the natural ecosystems is an unmanageable concern, even when addressed in some cases. As an example, after a lengthy public policy debate, sufficient concerns raised regarding the use of BPA in reusable water bottles led to an alternative product widely advertised as BPA-free. Yet the most common replacement for BPA is fluorene-9-bisphenol, or BHFP, which has its own set of concerns, maybe even worse than BPA but not identified as such. Was the remedy of the problem to replace BPA with a safer alternative or simply a BPA replacement? Without sufficient controls and understanding of plastic compounds across our global supply chain and a lack of consistency in the use of chemicals, especially as more manufacturing is happening in developing nations like China, should raise questions about the impact to us and our natural systems. As the use of plastics continues to spread in society, we must understand how the synthetic and natural supply chain systems are and are not linked to each other. An approach to chemical exposure is not able to manage the sheer volume of mass and permutations to identify what is a low or high dose, whether danger exists only in development stages or full life, and whether conclusions from animal studies can be translated into human impact.

Frankenstein 2.0

So there you have it: the plastic supply chain system, growing at almost a 9 percent clip over a century, and still growing both in volume, scope, complexity, and in geographical span, is running faster than our understanding and control over what is being produced, distributed, used, and disposed of across industry, society, and the environment. How can a global supply chain system track and manage a resin or product being produced in China or anywhere else in the world outside of its span of regulation and control? This is a question nobody is able to answer at this point in time, and a surefire sign that this is not the answer. Earlier in our use when it was more contained, fewer recipes, and at lower volumes, a closed-loop system could have been designed, not for the product itself, but for the overall supply chain system, but those days have come and gone, and now we are left with a runaway train of use without an understanding of how to stop it. Sure, we may understand how these base polymers work in concept, but we don't know how an unknown combination of undisclosed additives make plastic "plastic," or how all of these chemicals together impact an aggregate environmental or human system. Plastic is the most versatile material ever known to man, but it cannot exist in isolation of these natural systems; therefore, this definition of "versatility" needs to be more broadly defined to include plastic's use and reuse from the industrial to natural supply chain systems. If there are foreign, synthetic substances in our bodies due to plastic, and there is no doubt this is the case, we must better understand how these additives impact us, not just in isolation, as is being studied today, but in combination with other synthetics and at the genetic level that may provide a different picture of what's happening. Do we as the general public have a right to understand how synthetics that reside in us can impact our health? The answer is yes, and there should be no conflict in needing to understand these concerns with how we choose to live our lives in the modern, industrial world.

Every human and animal body tested in the past decade has shown chemicals that have leeched from plastics.[35] Even if you *choose* to live a plastic-free life, your body has chemicals in it related to plastic. You drink water, don't you? A 2017 study found 83 percent of tap water samples from around the world contained plastic pollutants and 94 percent of tap water in the United States.[36] If you doubt this study, or do not find it very concerning, what about chewing gum? Issues have been raised about a gum base called vinyl acetate as a possible carcinogen; the Canadian government at first classified it as hazardous and then concluded it isn't harmful to humans, perhaps due to pressure from the industry.[37] Who are we supposed to believe, and what should we do? These studies, and others, all done in isolation, may yield conclusions of no impact, but the human body is an integrated system, not just within itself but receiving inputs in many different ways, from many different sources. Is it causal, correlative, contributing, or just a mere coincidence that the rate of women's breast cancer has risen from 1 in 20 to 1 in 8, and newborns being born with 287 chemicals in their system? What about the invisible plastics, such as remnants from our washing machines, dishwashers, and facial scrubs that can empty into our water sources without being able to be detected or even properly filtrated? In a study conducted by the University of Gothenburg, one third of the 83 tested plastic products leeched toxic substances, including five out of thirteen intended for children.[38] The problem with these is they are very specific tests evaluating very specific uses of materials in very specific situations to question the impact on our human bodies. On the flip side, a test of what these chemicals are doing to us in totality would likely never pass the rigor of a scientific experiment due to an inability to control all the variables involved. It seems like a circular logic problem where it loops forward and back without any ability to resolve.

I believe the medical and research community is more informed than the general public, but insufficiently and without sufficient technologies to discern specific conclusions they can't assert clearly whether problems exist or not. Even with an improved knowledge of molecular genetics and nanoscience, there may not be sufficiently powerful microscopes to look closely at every corner of the globe, in every ocean gyre, at the farthest remote outpost void of human activity, or in the deepest trench on the planet. A 21st-century dynamic global supply chain system of interlocking supplies and demands beyond the eye of regulation and control must be matched by a 21st-century joint approach to medicine, science, and supply chains in order to understand what's happening on the ground and in our bodies. Today, we continue to be engaged in a 20th-century government model of environment, health, and safety to engage a health risk problem that cannot be analyzed using these crude methods.

In today's plastic and chemical supply chain system, the world is not so simple and well beyond the sophistication for this type of approach. Thousands of plastic and chemical combinations are introduced into the global market without sufficient review within our complex global supply chain systems. With a concurrent rise in cancers and other modern diseases seemingly connected somehow to industrialism, especially starting at earlier ages in development, an undisclosed, unclassified use of known toxins without any sort of regulation, market- or government-based, has been found to be associated in some way (causal, correlative, contributing, or coincidental) to trillions of product units produced every year in our supply chain systems.

According to Tracey Woodruff, a dramatic increase in chemical and, specifically, plastic production is also correlated to increases in conditions like attention deficit hyperactivity disorder (ADHD), autism, childhood cancers, diabetes, and obesity.[39] Is this merely coincidence, contributing, correlation, or causation? It would be foolish to consider any of this as a coincidence, and it might be rash to consider it as a causal one as well, but testing should seek to understand it without having to prove anything as a way of getting out of this circular reference where regulatory bodies are unable to prove something that cannot be proven. If it was our belief government regulation would keep us safe within a global supply-and-demand networked system, then it is our mistake to have assumed this. There needs to be a new paradigm of what it means for a product to be safe or not within a massively scaled and complex industrial system, especially how it links to our natural ecosystem and bodies. The big question is how to tackle this 800-pound gorilla.

Too Small or Big to See?

Is it feasible to put in place a management and control system into our massive and complex global supply chain system across nations, especially when even within them there are fragmented, dysfunctional political climates where science and reason could be under attack and ignored, and displaced with ideologies that revert to alternate facts and other trivialities? Maybe, but perhaps the first question is how to inform the public of these concerns. This will be a challenge unto itself among a population already facing too many other risks and concerns brought to their attention daily across a 24-hour news cycle. Most, if not a near unanimous percentage, of the population is unaware of the amount and variation of synthetic chemicals that exist in our bodies today not present just a few decades ago. There are movements of reform in the world, most notably in the United States, Canada, and Europe, where BPA has been banned for use in baby products, which just makes sense given the fragile nature of a newborn. And yet over the past decade, different studies conducted by the FDA have found BPA to be safe, something I have described in this chapter as such a generalized statement it means too little to us to be either concerned or not concerned, leaving us hanging as how to classify and understand these synthetics piling up in our lives and ecosystems. These types of studies are typically so specific as to isolate a nature of the problem, and then as a result are also equally limited in what the results tell us. Using 19th- or 20th-century definitions of toxicity and outdated technology as a matter solely of outdated science cannot peer through the lens of the 21st-century supply chain, meaning if the "dose makes the poison," then we are asking the wrong questions and, therefore, seeking the wrong solutions regarding plastic. Our global supply chain system is on its way to a production platform of 1 billion tons a year, much different from the 2 million tons that occurred in the 1950s, leading to new definitions of "dose" and "poison." As painful as it is to identify this moose on the table, it needs to be called out for the sake of our continued use. Our approach today can never succeed in proper research when it is researching the wrong questions.

Before you get depressed over this seemingly impossible research question to answer, consider if we fix what's wrong with our global supply chain system in order to manage plastic more responsibly without endangering it as economically viable, this question becomes less concerning. Nobody can deny a synthetic material introduced to these levels will not influence natural systems. It becomes a responsible next step for manufacturers and consumers to understand the links in order to become a part of the new process to view challenges and solutions in a manner to grow economies and use, not to reduce them. The limited view from the microscopes of scientists must be addressed as well because we must gain better insight into not only the invisible that can be brought into focus by science but also the view of the system that is actually too big for us to see as well, the perspective of the supply chain that is bigger than us as consumers or even of a single manufacturer and its role in the supply chain. That in itself is the question of Peak Plastic: what is too large and small to see that it becomes a threat beyond our ability to address? There remains a lot to be discovered after a century of fantastic growth to get bogged down into the minutiae of a specific question or study. The challenge in these concerns is not so much what can be seen and measured versus what cannot—and what cannot be seen is as much of what is practically invisible as is what is too big to see—but this behemoth supply chain system that has driven so much benefit from plastic.

I will present in the solutions chapters of this book that the spark to create these solutions will come from a point of understanding of us as stakeholders, of organic beings on an organic planet, who need to better understand how these materials are present everywhere in our lives are now so much in our bodies. This topic of plastic and synthetic materials is as unknown as it is important in our way of life, and it is impossible to separate it from being indispensable in modern society to being so damaging. Just the math itself of 2,000 new chemicals being introduced every year on top of 80,000 that already exist, with only a small minority of them (maybe as low as 6,000?) being classified and tested and an unknowable amount involved in plastic production[40] should lead to a desire of better understanding without throwing the baby out with the bathwater, as some would suggest, by banning them. There exists a debate of how we balance the protection of our environment, health, and safety and the need for economic growth across the planet. Certainly, I made a case that plastic is critical to the continued advancement of society as well as the continued reduction of poverty of millions and billions through economic growth. In the face of this critical nature of plastic in society, we need an improved understanding of this substance in all its forms that most, if not nearly all, of us do not understand. One thing is clear and we should all understand it: what we don't know has the potential to hurt us more than what we do. What is not known, and what we cannot see because it is too small and too large, is the cause of the problem, not the limits to our research and methods. Certainly, the medical and scientific research communities need to be more involved, but in the end, its basis is as a supply chain problem that must be resolved as a supply chain solution, and rests with us to think about as citizens and consumers. The supply chain must become the linkage between the industrial and synthetic and the ecosystem and natural. Once we view it in this manner, and begin to study the problems and solutions holistically rather than piecemeal, we will begin to get our arms around this and avoid this failure I am calling Peak Plastic. And we will solve it with our continuance of improving the lives of millions and billions through economic advancement.

References

1. National Academy of Sciences (U.S.). (1975). Materials and man's needs: Materials science and engineering: Supplementary report of the Committee on the Survey of Materials Science and Engineering. Washington, DC: National Academy of Sciences. Found at: https://www.nap.edu/read/10436/chapter/2#3 (accessed June 27, 2018).

2. National Research Council (U.S.). (1994). Polymer science and engineering: The shifting research frontiers. Washington, DC: National Academy Press.
3. O'Brien, Barbara. (May 2, 2018). Toxic plastics: Bakelite, the silent killer. *Sunny Ray*. Found at: http://www.sunnyray.org/Toxic-plastics-bakelite.htm (accessed June 27, 2018).
4. Imus, Deirdre. (July 27, 2015). Dangers of formaldehyde lurk in everyday products. *Fox News*. Found at: http://www.foxnews.com/health/2015/07/27/dangers-formaldehyde-lurk-in-everyday-products.html (accessed June 27, 2018).
5. Nicholson, Joseph L., and George R. Leighton. (August 1942). Plastics come of age. *Harper's Magazine*, 306.
6. Science History Institute. The history and future of plastics. Found at: https://www.chemheritage.org/the-history-and-future-of-plastics (accessed June 27, 2018).
7. Beall, Glenn. (April 9, 2009). By design: World War II, plastics, and NPE. *Plastics Today*. Found at: https://www.plasticstoday.com/content/design-world-war-ii-plastics-and-npe/27257907612254 (accessed June 27, 2018).
8. Weltwirtschaftsforum. (2016). The new plastics economy rethinking the future of plastics. https://www.weforum.org/reports/the-new-plastics-economy-rethinking-the-future-of-plastics.
9. Geyer, Roland, Jenna R. Jambeck, and Kara Lavender Law. (July 1, 2017). Production, use, and fate of all plastics ever made. *Science Advances* 3 (7): e1700782.
10. Westerhoff, P., P. Prapaipong, E. Shock, and A. Hillaireau. (February 1, 2008). Antimony leaching from polyethylene terephthalate (PET) plastic used for bottled drinking water. *Water Research* 42 (3): 551–556.
11. Sundar, Shyam, and Jaya Chakravarty. (2010). Antimony toxicity. *International Journal of Environmental Research and Public Health* 7 (12): 4267–4277; https://doi.org/10.3390/ijerph7124267.
12. Adeel. (February 15, 2016). Harmful effects of low quality PET bottles. Hassan Plas Packaging. Found at: http://hassanplas.com/harmful-effects-of-low-quality-pet-bottles/ (accessed June 27, 2018).
13. Wagner, M., M. P. Schlüsener, T. A. Ternes, and J. Oehlmann. (January 1, 2013). Identification of putative steroid receptor antagonists in bottled water: Combining bioassays and high-resolution mass spectrometry. *PloS One* 8: 8.
14. Recycling of HDPE bottles tops one billion pounds in 2012. American Chemistry Council. Found at: https://www.americanchemistry.com/Media/PressReleasesTranscripts/ACC-news-releases/Recycling-of-HDPE-Bottles-Tops-One-Billion-Pounds-in-2012.html (accessed June 27, 2018).
15. Hopewell, Jefferson, Robert Dvorak, and Edward Kosior. (2009). Plastics recycling: Challenges and opportunities. *Philosophical Transactions of the Royal Society B: Biological Sciences* 364 (1526): 2115–2126. Found at: https://www.ncbi.nlm.nih.gov/pmc/articles/PMC2873020/ (accessed June 27, 2018).

16. Yang, Chun Z., Stuart I. Yaniger, V. Craig Jordan, Daniel J. Klein, and George D. Bittner. (2011). Most plastic products release estrogenic chemicals: A potential health problem that can be solved. *Environmental Health Perspectives* 119 (7): 989–996.
17. Loyo-Rosales, J. E., G. C. Rosales-Rivera, A. M. Lynch, C. P. Rice, and A. Torrents. (January 1, 2004). Migration of nonylphenol from plastic containers to water and a milk surrogate. *Journal of Agricultural and Food Chemistry* 52 (7): 2016–2020.
18. EcologyCenter.org. Adverse health effects of plastic. Found at: https://ecologycenter.org/factsheets/adverse-health-effects-of-plastics/ (accessed June 27, 2018).
19. Walsh, Bill. (March 20, 2012). EPA reaffirms PVC's negative health impacts. *Healthy Building Network*. Found at: https://healthybuilding.net/news/2012/03/19/epa-reaffirms-pvcs-negative-health-impacts (accessed June 27, 2018).
20. Jaakkola, J. J. K., and T. L. Knight. (January 1, 2008). The role of exposure to phthalates from polyvinyl chloride products in the development of asthma and allergies: A systematic review and meta-analysis. *Environmental Health Perspectives* 116 (7): 845–853.
21. Occupational Health and Safety. (June 12, 2008). Study: PVC shower curtains potentially toxic. Found at: https://ohsonline.com/articles/2008/06/study-pvc-shower-curtains-potentially-toxic.aspx (accessed June 27, 2018).
22. Rogers, S. A. (July 6, 2011). Recycle Styrofoam cups: Is it possible? *Mother Nature Network*. Found at: https://www.mnn.com/money/green-workplace/stories/recycle-styrofoam-cups-is-it-possible (accessed June 27, 2018).
23. Food Service Packaging Institute. (2013). Foodservice packaging and...styrene. Found at: http://www.fpi.org/fpi/files/fpiLibraryData/DOCUMENTFILENAME/000000000287/Foodservice_Packaging_and_Styrene.pdf (accessed June 27, 2018).
24. Environmental Protection Agency. (January 2000). Styrene. Found at: https://www.epa.gov/sites/production/files/2016-09/documents/styrene.pdf (accessed June 27, 2018).
25. Gibbs, B. F., and C. N. Mulligan. (January 1, 1997). Styrene toxicity: An ecotoxicological assessment. *Ecotoxicology and Environmental Safety* 38 (3): 181–194.
26. Ahmad, Maqbool, and Ahmad S. Bajahlan. (January 1, 2007). Leaching of styrene and other aromatic compounds in drinking water from PS bottles. *Journal of Environmental Sciences* (China) 19 (4): 421–426.
27. Healtline. What is BPA and why is it bad for you? Found at: https://www.healthline.com/nutrition/what-is-bpa#section1 (accessed June 27, 2018).
28. Schug, T. T., A. Janesick, B. Blumberg, and J. J. Heindel. (November 1, 2011). Endocrine disrupting chemicals and disease susceptibility. *The Journal of Steroid Biochemistry and Molecular Biology* 127: 204–215.
29. Pinkerton, L. E., J. H. Yiin, R. D. Daniels, and K. W. Fent. (August 1, 2016). Mortality among workers exposed to toluene diisocyanate in the US

polyurethane foam industry: Update and exposure-response analyses: Mortality among TDI-exposed workers. *American Journal of Industrial Medicine* 59 (8): 630–643.
30. Hansen, Erik. (2013). Hazardous substances in plastic materials. Danish Technological Institute. Found at: http://www.miljodirektoratet.no/old/klif/publikasjoner/3017/ta3017.pdf (accessed June 27, 2018).
31. Dodson, R. E., L. J. Perovich, J. G. Brody, R. A. Rudel, A. Covaci, N. Van den Eede, A. C. Ionas, and A. C. Dirtu. (December 18, 2012). After the PBDE phase-out: A broad suite of flame retardants in repeat house dust samples from California. *Environmental Science and Technology* 46 (24): 13056–13066.
32. James, Maia. (January 14, 2013). How to avoid phthalates (even though you can't avoid phthalates). *Huffington Post*. Found at: https://www.huffingtonpost.com/maia-james/phthalates-health_b_2464248.html (accessed June 27, 2018).
33. Hamilton, Jon. (February 23, 2018). Plastic additive BPA not much of a threat, government study finds. *National Public Radio*. Found at: https://www.npr.org/sections/health-shots/2018/02/23/588356360/plastic-additive-bpa-not-much-of-a-threat-government-study-finds (accessed June 27, 2018).
34. Bushnik, Tracey. (2010). Lead and bisphenol A concentrations in the Canadian population. Ottawa: Statistics Canada.
35. Bergman, Å., et al. (2013). State of the science of endocrine disrupting chemicals 2012: Summary for decision-makers. Geneva: World Health Organization.
36. Carrington, Damian. (September 5, 2017). Plastic fibres found in tap water around the world, study reveals. *The Guardian*. Found at: https://www.theguardian.com/environment/2017/sep/06/plastic-fibres-found-tap-water-around-world-study-reveals (accessed June 27, 2018).
37. Thomas, Pat. (January 12, 2010). Behind the label: Chewing gum. *The Ecologist*. Found at: https://theecologist.org/2010/jan/12/behind-label-chewing-gum (accessed June 27, 2018).
38. Lithner, D., J. Damberg, G. Dave, and K. Larsson. (2009). Leachates from plastic consumer products: Screening for toxicity with Daphnia magna. *Chemosphere* 74 (9): 1195–1200.
39. Miller, Molly. (June 22, 2017). Toxic exposure: Chemicals are in our water, food, air and furniture. *UCSF News*. Found at: https://www.ucsf.edu/news/2017/06/407416/toxic-exposure-chemicals-are-our-water-food-air-and-furniture (accessed June 27, 2018).
40. Scialla, Mark. (June 22, 2016). It could take centuries for EPA to test all the unregulated chemicals under a new landmark bill. *PBS*. Found at: https://www.pbs.org/newshour/science/it-could-take-centuries-for-epa-to-test-all-the-unregulated-chemicals-under-a-new-landmark-bill (accessed June 27, 2018).

CHAPTER THREE

The Waste-Driven Supply Chain

Fake Plastic Trees

During World War II, consumption was under strict control in the United States; citizens were given coupon books that limited what, when, and how much of an item could be purchased. The War Production Board regulated all materials; civilian men and women wore victory suits to cut down on fabric, and silk stockings were banned altogether. As the war ended, there became a new set of concerns: rather than sacrifice and a rationing of resources to support a war effort, it was the enormous war-manufacturing capacity without sufficient demand that worried the government leaders to the risk of yet another economic depression to follow. An enormous industrial manufacturing engine was built for war, and now it was to be idled as the consequence of peacetime. Plastic was the material that helped supply the war, and it also seemed to be the miracle material to win the peace. This was the moment when plastic became more than a material but rather an economic engine and eventually a cultural icon that grew so substantially. As the global supply chain system emerged and grew with the help of big-box supercenters like Walmart, plastic products became cheaper, lighter, stronger, more functional, and practically unlimited. How would it have been possible for this post–World War II consumption-based economy to be shaped without plastic?

This cultural element of plastic is important to emphasize: in 1995, the alternative rock band Radiohead released the song *Fake Plastic Trees*, what would become a metaphor of our culture of materialism and how it has consumed us and our values. This meme can be paraphrased as nothing today is too valuable, including our natural resources and ourselves, and there's a supercenter down the road or a website within a few clicks to satisfy all your wonted needs. In contrast, I still remember stories my grandmother recollected from World War II when nearly everything she and her family owned would have to be repaired, and practically nothing was discarded in an effort to win the war and be called upon as a reflection of character. In extreme contrast, my children's generation has been taught since birth that materialism is most important, with nothing worth saving, and in some respects, the same lesson can be translated in regard to human relationships and other intangibles. This is a battle of ideas in the definitions of value and waste, with the latter being driven by our supply chain systems while the former based more on traditions, values, and a connection to nature. In biology, waste is a residue of value to be reused as a part of an overall closed-loop system, a resource, a meaningful aspect of how materials are consumed and reused where practically nothing is left as trash. This is a result of ancient synergies perfected over time that would ensure a closed-loop system not as a matter of sustainability but rather of growth. How can life grow if there's no waste to be reused? For most of human history, man has also followed these ancient codes as both a matter of biological and cultural evolution. And yet, just a few centuries ago, these rules that were passed down from nature to society were broken by a new covenant enabled by plastic, a material deemed to be separate and bulletproof from the laws of nature. It makes perfect sense, in a strange way, that the use of synthetics that is nature-resistant can extend the food chain, a necessary innovation in a world of 7 billion humans and growing. Yet, in totality, it has extended life in unnatural ways, proliferating waste instead of reusing it, and is not sustainable in the long run.

Many, if not most, of today's supply chains thrive on bad waste (not to be reused) rather than good waste. Consumer markets grow as a function of waste as does our back-end waste-management system that whisks it all away without much consumer obligation and significant profit. I consider myself to be the transition generation between my grandmother, who was raised in a culture of leaving no waste as a matter of civic responsibility and market reality, and my daughters, who have been raised in a completely opposite paradigm. I remember going with my father to the local landfill as a young boy where an inspection clerk would carefully review our items to be discarded, and the costs were rather significant. This is in contrast of today where practically anything you put out on your curbside to be discarded will be accepted by the waste collectors, especially if you tip them during the Christmas season! Just in plastic alone, there is an estimated 6–8 billion tons produced since its onset, with only 9 percent recycled, 12 percent incinerated, and the remaining 79 percent accumulating in landfills, or worse, in the natural environment.[1] Given these results, there is no question the root cause of this problem of Peak Plastic is a waste-driven supply chain that is not only in place to produce one-way disposable materials but also to unintentionally drive these behaviors through all agents of the supply chain system. And yet today, decades after we have put in place the business systems and innovations to drive these systems to continuous, one-way, synthetic growth, we are beginning to learn how this substance breaks down so unnaturally to have great potential harm to our oceans and own bodies, perhaps causing us to take pause of what we do next. Yet because it embodies our culture, it won't be as easy as to simply turn it off as a matter of new principles or obligations. Man has become the only creature to ever be able to foster a culture of bad waste, and such a liberation from nature won't be so easy to relent.

We need to understand the deep cultural elements of this plastic problem, our waste-driven supply chain systems, as where the corrections must begin. Today, this culture has become a biological problem as well, embalmed in the sea ice in the Arctic, our water systems, and even our own bodies. Such problems may appear to be out of sight for the most part in the United States, or just a nuisance, but in Asia, it is leading to perhaps the greatest planetary crisis in our human history, having the potential to surpass the impact of climate change as well. Asia's huge, largely unregulated growing megacities exist without sufficient municipal planning that is leading to an epidemic of waste, much of which consists of plastic and other synthetic materials. Asia is also the industrial factory for our planet and, by default, the plastic capital of the world, transforming such ancient cultures as China, India, Vietnam, Indonesia, and the Philippines into hazmat regions in a matter of a few decades. This has become the Faustian bargain: that a global supply chain of waste will pull hundreds of millions out of poverty, which it is doing but at a terrible price; peasants are migrating to these megacities to become factory workers and, unfortunately, waste pickers. An eradication of poverty in Asia, Africa, and Latin and South America through plastic waste? It's just another plastic paradox of how our current state economies and their supply chain systems work, not just in a trade-off of how waste breaks down the environment but also in how it leads to more of a circular pattern in creating poverty as yet another unexpected by-product.

Of the Slum and the Sea

Did you know half of all the plastic ever made in human history has been produced in the last 13 years?[2] Do the math: half of the plastic produced in history has occurred in the last 13 years and is growing, and nearly 80 percent of all plastic ever been produced is now discarded as trash and is growing. The proliferation of these global supply chains is driving a mass migration of hundreds of millions of Asians and others from poor, rural communities to these chaotically, unstructured megacities, the super-urban population centers of over 10 million people and emerging economic growth that are today's Peak Plastic capitals of the world. In these regions, 70 percent of the growth is unplanned and one third of the inhabitants live in slums or informal communities that lack basic municipal services like sufficient waste disposal practices.[3] Further exacerbating these Peak Plastic epicenters is that many of these megacities are housed on or within 50 kilometers from the ocean coast, a factor that is the largest determinant for the source of marine-based plastic pollution. Two of the largest plastic pollution contributors are Indonesia and the Philippines, which have 74 percent and 83 percent of their populations living in these coastal regions, respectively.[3] This is the Peak Plastic *perfect storm*: the fastest-growing population and consumer regions in the world, possessing a population with higher percentages of poverty, undergoing a mass migration from rural areas to dense super-cities situated on the oceans and other waterways, forged to unregulated mass manufacturing centers with poor waste management practices. Global plastic production has tripled over the past 20 years, and Southeast Asia represents 20 percent of global output and is growing.[3] This is already, no doubt, a planetary crisis. The fragility of this situation is almost impossible to overstate; there are a quarter of a billion people in East Asia living in slums, and as their conditions continue to deteriorate, a greater crises will emerge and expand geographically. Regarding Asia as the manufacturing center in the world, in 2012 a large typhoon hit Hong Kong and caused damage to containers in their port, leading to 150 tons of plastic pellets spilling into the South China Sea.[4] As instances like this occur and these plastic pellets wash ashore as "Hong Kong snow" or are set out to sea as a synthetic disruption to our natural ecosystem, Asia becomes the epicenter for Peak Plastic, perhaps a vision to what our future may be in Western societies.

In a largely ceremonial yet important announcement at its environmental summit held in Nairobi, Kenya, in December 2017, the United Nations made a resolution that the world needs to combat plastic waste from entering the oceans. Under this proposal, the UN would establish a task force to advise on combatting what the UN Oceans Chief is calling a "planetary crisis."[5] Well intended, no doubt; a very important communiqué but one of no binding authority and certainly little to any influence over a global supply chain system that relies on plastic as a model of economic growth, especially for the developing world. Messages like these will be important moving forward but will not lead to any substantive measures to change the tide (no pun intended); in my solutions chapters, I will present what will need to be done to face such a challenging case for change. There are also well-intended entrepreneurs who are focusing on the issue of plastic waste, yet many of these proposals are focused on the symptoms and not the root cause problems. Take The Ocean Cleanup project, an NGO initiative based in the Netherlands that deploys mobile devices into the ocean to clean up three tons of plastic debris a week per device. There is so much good to say about this project, including the ambition of an entrepreneur at the age of 17 who combined his concern over plastic waste in the ocean and his love of technology to develop these systems that are about to be tested off the coast of California for an eventual launch to the Great Pacific Garbage Patch to clean up that environmental problem. Yet what is the opportunity cost of such an initiative, and are there more suitable alternatives to addressing the root cause rather than the symptoms? According to some experts, placing this unproven technology in the middle of the ocean to clean up debris that imprinted itself hundreds if not thousands of miles along the way is an expensive solution for a smaller part of the problem (only 1 percent of debris is on the surface).[6] This is no slight to young entrepreneur Boyan Slat of the Netherlands, but these types of niche solutions cannot solve the overall systems problem (supply chain and natural) of plastic; how will our oceans be cleaned if there's a constant slime of plastic and its additives as they accumulate thousands of miles to the five gyres of the ocean? My concern is that grand proclamations and technological gadgets sponsored by celebrity investors may become so pronounced they suck all the oxygen from out of the room, leaving no time or resources for more fundamental, root-cause, common-sense solutions to the problems. To get there, we must first understand the problem starts not in consumers littering or not recycling but rather in the culture of a supply chain system driven by waste. In Thailand, one of the fastest growing plastic markets in the world, the government has announced a 20-year plan to address this problem of plastic waste, a strategy without the level of urgency required, according to the UN, to address this planetary crisis.[7] These and other incremental solutions can only mitigate and not solve for these problems, and a full-cycle analysis should be conducted to determine whether such plans can even keep up with the pace of growth and damage happening today. Just one plastic commodity, the plastic bottle, is produced worldwide at 20,000 bottles a second, with approximately 7 percent of them turned back into bottles and the remaining 93 percent left as a growing problem.[8] The reality of the math of these incremental solutions added to a single-digit recycling rate does not add up against a plastic supply chain system that will double in production in 20 years and quadruple by 2050. Nothing against UN proclamations, government policies limited in scope, and incremental innovators, but these efforts can be seen as distractions from the root-cause problems necessary to be addressed on a timeline of Peak Plastic.

Our world is racing toward a population of 10 billion people through a disproportional growth from the least developed pockets of the world. Because plastic is so cheap, versatile, and bountiful, it will surely become the go-to material for the poorest of the poor, where scarcer and more expensive resources will not attend. In these developing nations, the poor will begin to emerge as consumers while the poorest will play the role of an informal waste management system, employing tens of millions of waste pickers at below-poverty wage levels in these developing cities. According to a 2010 UN report, waste pickers can collect up to 50–100 percent of waste collection in a developing nation without an impact to the municipalities' budget; as a stretch to the term, these individuals are called "green workers," relying on the belief that something is always better than nothing.[9] Imagine the travesty of a poor slum family living in an unregulated landfill classified as green workers! Yet sometimes these waste pickers become a part of a cooperative, offering the workers a guaranteed income, but often with strings attached. These alleged informal waste management systems can both protect and take advantage of the poorest of the poor, turning into enterprise opportunities for some to make more than others. Therefore, a development of these informal supply chains cannot be glorified beyond being an occupation of last resort, especially given the lack of a valid financial model for these collectibles, as I have mentioned as the root-cause problem for plastic. Any way you slice it, it's a crisis.

The waste picker occupation is one of last resort, and these individuals are some of the most humiliated and disrespected people on the planet. At this moment, there are millions of people who live not just near a landfill but actually on the site, as many as tens of thousands at a single location, with a life expectancy of 35 given the rampant spread of disease such as pneumonia and tuberculosis. It's difficult for us to imagine how bad of a life this must be, even if we use our worst-case scenarios to attempt to consider the sights, smells, despair, and lack of dignity related to it, and yet it is a critical role in a developing community. I have direct insight to this problem in the nation of Namibia, a small country in southwest Africa that has struggled over centuries as a colonial subject of Germany and then Great Britain and South Africa prior to gaining its independence in 1990. Despite its development into a flourishing young democratic republic system, it is one of the most unequal nations in the world as it seeks to untangle centuries of apartheid leading to the largest percentage of the population possessing just a sliver of the nation's wealth and opportunity. In Namibia, a nation that relies heavily on its neighbor South Africa to act as its de facto supply chain system, the fetch rate for a kilogram of used plastic PET bottles is N$0.39, or slightly less than US$0.03 per kilogram. As a result of Namibia getting its water and soft drinks served in plastic bottles from neighboring South Africa, the fetch rate of used PET in the former is five times lower than the latter, an indication of the importance of the supply chain system in the recycling/reuse process. On a typical day, a waste picker from Namibia will collect enough PET bottles to fill five 5-kilogram bags, or 25 kilograms a day. At its fetch rate, this translates to a daily wage of US$0.74, which is significantly below the latest UN definition of world poverty of US$1.95. Collecting over 800 PET bottles in a day (831) could take an entire day, yet the work does not allow the individual to escape poverty conditions. When I asked my Namibian colleague how this can be, he quizzically answered that there really is no other choice for too many. In this country, 22.4 percent of Namibians are living below the poverty line and there is an almost 30 percent unemployment rate despite a growing economy.[10] The waste picker in South Africa who collects this amount of used beverage containers will earn approximately double the official poverty rate, an amount of US$3.77, a difficult fate but much better than that in Namibia.[11]

In the poorest regions of the world where population growth is the heaviest, there is both an insufficient waste management infrastructure in place and a supply chain system unable to foster a balanced flow between the inflow of bottled water and other products and the outflow of plastic waste. This leads to a vicious cycle between human well-being and environmental stability, with the degradation of humans living under unspeakable conditions wreaking havoc on our oceans and vice versa. In this circumstance, we should be able to acknowledge how an impact to the environment is an afterthought to these vicious cycles of poverty having no apparent remedy. Unfortunately, our supply chain systems have inextricably linked a choice of seeking to remedy poverty through consumption versus such a deadly impact on our oceans, an impossible choice to be forced to make. Of course, an NGO organization can make an emotional case of dead seabirds full of plastic, but there are many unpublicized cases of children dying at an early age of disease while being forced to pick through trash sites in the midst of feral animals. Much of this we don't want to know about, and yet if we make the effort to consider what's happening and we define the problem properly, we will see it as an extension of a broken global supply chain system that serves billions efficiently in our modern day society. It is the same supply chain that created the problem that can fix it.

The Black Box of Trash

Our oceans comprise 70 percent of Earth's area, and yet we know so little about them. Every year, we pull 200 billion pounds of food from them that is critically important to the poorest regions of the world where nutrition is at a premium.[12] We also gain 50 percent of our oxygen from it, a growing percentage of the water we drink via salinization, and an ecosystem critical to the regulation of the planet's climate, a very important responsibility.[12] What most of us do not understand is we have explored less than 5 or 10 percent of our planet as a result of our lack of understanding of our oceans.[13] For example, there are underwater waterfalls, lakes below the ocean, mountain ranges, valleys, an immeasurable number of species undiscovered, and even black smoker vents 2,000 meters below the surface that are a paradigm shift away from our notion that the sun is our sole energy source on Earth. We know so little about this piece of our planet that is so important to our own well-being, and as a result, it has become a trash receptacle under the perspective of "out of sight, out of mind." Every year, humans intentionally and unintentionally dump 19 billion tons of garbage into the oceans with much of it as plastic, and this number is expected to double by 2025.[14] This currently tallies to tens of trillions of pieces of plastic in the oceans equaling 46,000 to 100,000 pieces of plastic per square mile, leading to two thirds of the fish population ingesting it due to mistaking it for food.[15] This becomes a vicious cycle for humans and the fish population: an increase in plastic in the ocean both damages the health of the fish population and humans' use of seafood. Some scientists have predicted by the year 2050, there will be more plastic in the ocean than fish.[16]

How such a catastrophe occurs is truly a perfect storm of error: a seemingly harmless material of plastic that is misunderstood being dumped into oceans that are barely understood themselves. Beyond an image of plastic perceived as inert and uneventful, the impact of plastic to our oceans is often difficult to properly measure or draw thoughtful conclusions from. And then there's the problem of who is responsible for protecting the water thousands of miles out from any national territory and five miles deep, which belongs to none and all of us at the same time? Will the developed world consider ourselves responsible for fixing the plastic pollution problem when 75 percent of it is caused by the five developing nations who are some of our largest global factories?[17] Depending upon the source of your information, this mismanagement of plastic is affecting from 600 to 1,300 species, many of them in our oceans; this is our best estimate, and the numbers of species impacted is probably much greater given our lack of knowledge of the ocean. What policy should we undertake to address this; is it a sovereign problem for us to worry only about our shores, everyone's, and the deepest parts of the ocean? The list of marine animals that confuse plastic as food, such as a leatherback sea turtle believing a plastic bag to be a jellyfish, is perhaps more numerous than those that do not. A recent photo taken by photographer Justin Hofman in Borneo of a seahorse using a littered cotton swab for balance went viral, evidence of this dangerous and growing interaction between wildlife and synthetic materials in the oceans. On the sea surface, plastic is highly resistant to degradation, perceived as harmlessly sinking to the bottom of the ocean, but once it is there, it will leech and break down under higher pressure, releasing toxic chemicals as a result. In the deepest part of our Earth, the Mariana Trench that stretches 10,000 meters deep in the Pacific Ocean, small crustaceans were captured by robots and found to be 50 times more polluted than crabs that have survived in some of the most polluted rivers in China, which demonstrates the dangerous impact we are having on our oceans and just beginning to realize.[18]

Ocean plastic is harmful in many ways; from one study, 25 percent of fish purchased from fish markets in Indonesia and the United States had plastic in their guts, and in a 2014 study, it was found the average European seafood consumer could be eating up to 11,000 micro-plastic pieces a year.[19] Over a persistent and long degradation process, perhaps as long as 50 years, these materials disperse and bio-accumulate, providing marine life plenty of opportunity to be impacted and ingest these synthetics. While the impact of macro-plastics on sea life is often fatal through choking or internal damage, micro-plastics can be ingested as food and have a much more harmful impact to the food chain. Starting with zooplankton, the smallest of ocean species are consumed by larger fish also consuming plastic, and it works its way up the food chain directly and indirectly to us. Eventually, plastic either is consumed by marine life, sinks to the ocean bottom and wreaks more havoc, or is taken by the currents to one of the five ocean gyres located in the North and South Atlantic and Pacific Oceans. As a result of winds and ocean currents, these gyres naturally collect pollutants, mostly plastic. The most noted of these gyres is the Great Pacific Garbage Patch, believed to be at least 270,000 square miles but recently determined to be four to sixteen times larger than once projected.[20] In this region, the Scripps Institution of Oceanography found evidence of plastic debris in 100 consecutive samples, demonstrating how great of a problem this is for our oceans.[21] On the surface there is an estimated 270,000 tons of plastic floating on the ocean, and below there are pieces smaller than 5 millimeters in size that accounts for 92 percent of the ocean plastic.[22]

The "Nano" Catastrophe

Did you know 92 percent of the problem of plastic in our oceans is unseen and largely undetectable or treatable? Most of us would believe what we see as an eyesore is the problem, but as the saying goes, there's more to it than meets the eye. Here's the statistics: 80 percent of the plastic in the ocean is commenced from the land (the other being dumped while in the ocean), with 15–31 percent released into the ocean already as micro-plastics, with two thirds of it remnants of fibers from washing machines and tire abrasion from the roads.[23] In fact, a single load of laundry you do at home can release more than 700,000 micro-plastic fibers into the environment, with trillions of pieces entering our oceans on an annual basis.[23] These numbers present a "nano" catastrophe, a problem invisible to us, unlike smokestacks and burning rivers and therefore without a public outcry to do something about it, no "Crying Indian" as it was called in the 1970s. Calling this phenomenon a nano catastrophe may not accurately convey the danger of the invisible, however—the problem impacting our health and our environment is too small to detect, so too small to pay attention to and correct. Solving the threat of an asteroid from outer space seems daunting yet doable because it is one massive rock capable of dooming civilization, similar to how it offed the dinosaurs; we get a decade of advanced notice and give it a lot of attention. In contrast, these trillions of particles of waste cannot be isolated or identified, preventing any notice or coronation toward the problem! Micro- and nanoplastics are generated in such a large array of industrial and domestic products, they often act as an invisible sponge to absorb toxins, pesticides, and other dangerous chemicals on the path to our oceans. This plastic debris less than 0.5 millimeters in diameter should be considered a major concern given the lack of research capability to understand it at this point. And it is not even the base plastic polymer in even these micro sizes that is of greatest concern but rather the additives and other injurious compounds that can hitch a ride to these virtually indestructible materials as they make their way to the depths of the oceans or to these five gyres. On this journey, microorganisms are likely ingesting these chemicals that are in turn working their way up our food chains through a process called "biomagnification" in a completely undetected manner. Studies have found plastic fragments circulating in the blood of the mussels we eat, a disturbing fact not listed on the menu where they highlight the origin of this expensive appetizer.[24] All the pictures and statistics of dead sea birds, whales, and dolphins illustrate a problem, including 70 percent of dead dolphins found stranded on the beach have ingested considerable amounts of macro-plastic, but it is what we can't see or measure that is the greater problem.[25] Yes, plastic bottles in the ocean are bad, but it is those exfoliation beads from facial scrubs, tire remnants, fibers from your laundry, micro-bits from your dish detergent pods, and the breakdown of these larger pieces of plastic over months, years, and decades that are perhaps the greater threat to us and the ocean. As our fish populations continue to crash, with tuna and mackerel populations declining 75 percent in the last 40 years and the remaining 25 percent being potentially toxic, we must begin to have more honest conversations as to whether our current behaviors that consider our oceans as these black boxes of unintentional misunderstanding are really okay.[26] Unlike related to our own health, researchers have been able to find clear causal relationships between plastic pollution and marine life welfare, meaning the problem is largely indisputable. There is so much we need to learn about these deep dark black boxes that comprise 70 percent of the world, especially given how important they are to us and how much of this synthetic stuff we are carelessly dumping into them.

A "Glocal" (Global and Local) Problem

Is this a planetary crisis or just a bunch of local problems? Stories of Hong Kong snow, tens of millions of people living thousands of miles away on mega trash heaps to survive, falling fish populations, and a growing balance of toxins in oceans and aquafarms that are the source of much of our seafood is what's happening locally and impacting everyone globally. In these places, a vicious cycle is happening between poverty and environmental damage that places more pressure on both as well as the rest of the planet: both in our homes at our dinner table and in the great oceans where all of humanity is impacted. The densely populated, poor regions of Asia and Africa provide irrefutable and indescribable examples that express themselves locally and globally, and yet on our homelands, these crises appear as well. Flint, Michigan is a city that has been dealing with a municipal water crisis now for four years, leading to a vicious cycle between a lack of clean water, economic opportunity, and environmental damage. In 2016, two years after the start of Flint's water crisis, resident and film documentarian Michael Moore made a plea for people to stop sending bottled water to Flint, because if the need for people is 200 bottles for 102,000 citizens, that's over 20 million bottles in a year that would require a new landfill, making the problem even worse; unfortunately, the problem still persists today, two years after Moore's plea.[27] In 2017, in an American territory, Puerto Rico, there was also a water crisis with nearly half of the island's residents of approximately 1.5 million people not having access to clean drinking water and having to use bottled water as a result. Prior to Hurricane Maria, 70 percent of the population drank unsafe water, and the water system today is in a state of crisis that will take years to fix.[28] Trying to address the problem of a lack of clean water leads to a trade-off for another problem, the use of bottled water as an alternative that increases the plastic waste impact to the environment, which is an especially egregious problem for an island municipality. In areas like Flint, Puerto Rico, and in other communities, Americans are either without clean water or facing the threat of losing it; more than one third of all Americans are at risk of losing affordable drinking water.[29] Is this growing problem in a developed nation like the United States correlated to a growth in the use of commercial bottled water, a greater use of plastic, or simply a matter of a crumbling infrastructure system? Regardless of whether this is mere coincidence or not, if there are growing challenges in clean water in this nation, and it's being mitigated through a greater dependency on commercial bottled water, a vicious cycle between municipal and environmental problems will grow, as they exist in Asia and other developing areas of the world. While this problem of Peak Plastic in the United States may not be as severe as it stands today in Asia, there is evidence we are beginning to see the signs of it beginning to take root if we look close enough. It's a fallacy to believe this is only a problem in the developing world, and as we know, it becomes a self-fulfilling and cumulative problem within our one planet, especially through our rivers and oceans.

Even if municipal water systems were not at risk of failing in the United States and other developed nations, what's happening in the developing nations will become a growing crisis with a significant yet unheralded impact on the rest of the world. In our fluid ecosystem and global supply chain system, problems cannot remain localized; the extent of environmental decay in Asia, for example, cannot be contained to itself, and the global supply chain is an unleashing of globalization to every end of the planet. In this model, America's consumption of shrimp has increased since the early 1990s as a result of the global supply chain of lower cost production in China.[30] Today, China accounts for 60 percent of the global aquaculture business, a $90 billion market.[31] As a result of our global supply chain system, nearly all of the U.S. shrimp supply is imported (94 percent), the FDA inspected between 0.7–3.7 percent of these exports, and a *Consumer Reports* study found 60 percent of frozen shrimp to be contaminated with bacteria.[32] In a global supply chain system of aquaculture and plastic, problems converge, both literally and figuratively; a polluted waterway or other environmental challenges can impact other models, such as our global seafood model that is largely from some of the same regions of the world as industrial manufacturing and the most polluted environments. In this global seafood factory farm model, there have been significant concerns regarding bacteria, antibiotics, and toxin traces in imported seafood. Dr. Martin Blaser, a professor of microbiology from New York University who has chaired advisory panels for Barack Obama, has raised concerns regarding antibiotics and bacteria in our food, particularly food imported from nations with questionable standards. A global food supply chain system that seeks to meet our needs of higher volumes of food at lower prices runs the risk of serving us from these areas with questionable water sourcing contaminated by these chemicals and plastics. You may think an ocean is between you and the problem, but the global supply chain system brings the problem to you in many ways, such as your dinner table—and the ocean itself, a fluid system, is the problem.

No doubt, this interconnected global supply chain has been responsible for an unprecedented degree of peace and economic prosperity over the last few decades, particularly to the largest and most populated region of the world, Asia. Yet in our closed ecological environment and a world population of 7.5 billion that is on its way in this century to 10 billion, this is an unsustainable model beginning to rear its ugly head across the world and in what used to be our pristine oceans, which are so crucial to the health of our planet. Regardless of politics, we are a part of an interconnected planetary system of economics and the environment whether we like it or not. The interdependency of a global supply chain system that grows, manufactures, and ships from some of the most contaminated places on earth, the epicenters of Peak Plastic, began with American culture, of this throwaway society to enable consumerism to ensure economic growth and to pull people out of poverty and into a new way of life. In the United States, this culture of a throwaway society commenced after World War II and has expanded to every reach of the planet, and it has become unstoppable in these megacities of unmanageable growth. Even if intergovernmental bodies like the United Nations had effective methods and channels of change, which they do not, they cannot hold sway over these unplanned and informal megacities that have become linchpins to this supply chain network. And yet, these are the opening acts for the future of Peak Plastic, and we as a planet and supply chain are almost beyond the point of crisis. If it is too late for the United Nations to intervene, if this body ever had the power to do so to any material extent, what is the hope of curtailing this runaway plastic supply chain system that is spilling into our natural systems in such a catastrophic manner? It will not be up to its intergovernmental body, or the individual government bodies of each sovereign nation, or to us as consumers. Then whom? Not nifty entrepreneurs who develop gadgets that can only make a negligible difference. It will be up to the same global supply chain system that is bringing us this problem as a function of eradicating world poverty but perhaps not the same actors who pursue economics without environmental sustainability. It will be the leaders of disruptive innovation through a 21st-century approach of science, technology, and supply chain—and if so, we still have a chance of solving for Peak Plastic. But first, we must understand it is upon us, and there is not a lot of time to get started on it—so let's go!

References

1. Geyer, Roland, Jenna R. Jambeck, and Kara Lavender Law. (2017). Production, use, and fate of all plastics ever made. *Science Advances* 3 (7): e1700782.
2. Zhang, Sarah. (July 19, 2017). Half of all plastic that has ever existed was made in the past 13 years. *The Atlantic*. Found at: https://www.theatlantic.com/science/archive/2017/07/plastic-age/533955/ (accessed June 29, 2018).
3. Scott, Roy. (April 2015). Managing marine plastic pollution: Policy initiatives to address wayward waste. *Environmental Health Perspectives* 123 (4).
4. Petz, Bob. (August 10, 2012). "Plastic snow" nurdles wash ashore in Hong Kong. *Ecology Today*. Found at: http://www.ecology.com/2012/08/10/plastic-snow-washes-ashore-hong/ (accessed June 29, 2018).
5. Harrabin, Roger. (December 5, 2017). Ocean plastic a "planetary crisis"—UN. *BBC News*. Found at: http://www.bbc.com/news/science-environment-42225915 (accessed June 29, 2018).
6. Solletty, Marion. (January 16, 2018). Scientists skeptical over quick fix to ocean plastics problem. *Politico*. Found at: https://www.politico.eu/article/scientists-skeptical-over-quick-fix-to-ocean-plastics-problem/ (accessed June 29, 2018).
7. Corben, Ron. (June 22, 2017). Asia's booming plastics industry prompts ocean pollution fears. *Voice of America*. Found at: https://www.voanews.com/a/asia-plastics-industry/3911586.html (accessed June 29, 2018).

8. Laville, Sandra, and Matthew Taylor. (June 28, 2017). A million bottles a minute: World's plastic binge "as bad as climate change." *The Guardian*. Found at: https://www.theguardian.com/environment/2017/jun/28/a-million-a-minute-worlds-plastic-bottle-binge-as-dangerous-as-climate-change (accessed June 29, 2018).
9. Dias, Sonia. (November 27, 2012). Not to be taken for granted: What informal waste pickers offer the urban economy. *The Global Urbanist*. Found at: http://globalurbanist.com/2012/11/27/waste-pickers (accessed June 29, 2018).
10. Republic of Namibia National Planning Commission. (2015). Poverty and deprivation in Namibia. Found at: http://www.na.undp.org/content/namibia/en/home/library/poverty/nimdpovmao2015.html (accessed June 29, 2018).
11. Blaauw, P. F., J. M. M. Viljoen, C. J. Schenck, and E. C. Swart. (April 2015). To "spot" and "point": Managing waste pickers' access to landfill waste in the North-West Province. *AfricaGrowth Agenda* 4 (6): 18–21.
12. Rauch, Paul. (2016). Forsaken Earth: The ongoing mass extinction. Raleigh, NC: Lulu Press.
13. Friedersdorf, Conor. (June 28, 2013). Unfathomable: How much we don't know about the ocean. *The Atlantic*. Found at: https://www.theatlantic.com/technology/archive/2013/06/unfathomable-how-much-we-dont-know-about-the-ocean/277328/ (accessed June 29, 2018).
14. Mosbergen, Dominique. (May 12, 2017). The oceans are drowning in plastic—and no one's paying attention. *Huffington Post*. Found at: https://www.huffingtonpost.com/entry/plastic-waste-oceans_us_58fed37be4b0c46f0781d426 (accessed June 29, 2018).
15. Ocean Crusaders. Plastic ain't so fantastic. Found at: http://oceancrusaders.org/plastic-crusades/plastic-statistics/ (accessed June 29, 2018).
16. Kaplan, Sarah. (January 20, 2016). By 2050, there will be more plastic than fish in the world's oceans, study says. *Washington Post*. Found at: https://www.washingtonpost.com/news/morning-mix/wp/2016/01/20/by-2050-there-will-be-more-plastic-than-fish-in-the-worlds-oceans-study-says/?utm_term=.38522b2a314f (accessed June 29, 2018).
17. Winn, Patrick. (January 13, 2016). Five countries dump more plastic into the oceans than the rest of the world combined. *Public Radio International*. Found at: https://www.pri.org/stories/2016-01-13/5-countries-dump-more-plastic-oceans-rest-world-combined (accessed June 29, 2018).
18. Carrington, Damian. (February 13, 2017). "Extraordinary" levels of pollutants found in 10km deep Mariana trench. *The Guardian*. Found at: https://www.theguardian.com/environment/2017/feb/13/extraordinary-levels-of-toxic-pollution-found-in-10km-deep-mariana-trench (accessed June 29, 2018).
19. Kerlin, Kat. (September 24, 2015). Plastic for dinner: A quarter of fish sold at markets contain human-made debris. UC Davis. Found at: https://www.ucdavis.edu/news/plastic-dinner-quarter-fish-sold-markets-contain-human-made-debris/ (accessed June 29, 2018).

20. Kennedy, Merrit, and Christopher Joyce. (March 22, 2018). The trash patch in the Pacific is many times bigger than we thought. *National Public Radio.* Found at: https://www.npr.org/sections/thetwo-way/2018/03/22/596142560/the-trash-patch-in-the-pacific-is-many-times-bigger-than-we-thought (accessed June 29, 2018).
21. Aguilera, Mario. (September 1, 2009). Voyage to the plastic vortex. Scripps Institution of Oceanography. Found at: https://scripps.ucsd.edu/news/voyage-plastic-vortex (accessed June 29, 2018).
22. Eriksen, Marcus, Laurent C. M. Lebreton, Henry S. Carson, Martin Thiel, Charles J. Moore, Jose C. Borerro, Francois Galgani, Peter G. Ryan, and Julia Reisser. (2014). Plastic pollution in the world's oceans: More than 5 trillion plastic pieces weighing over 250,000 tons afloat at sea. *PloS One* 9 (12): e111913.
23. Boucher, Julien, and Damien Friot. (2017). Primary microplastics in the oceans: A global evaluation of sources. Gland, Switzerland: International Union for Conservation of Nature and Natural Resources.
24. Browne, M. A., A. Dissanayake, T. S. Galloway, D. M. Lowe, and R. C. Thompson. (2008). Ingested microscopic plastic translocates to the circulatory system of the mussel, Mytilus edulis (L). *Environmental Science and Technology* 42 (13): 5026–5031.
25. Baulch, Sarah. (May 24, 2013). The shocking impacts of plastic pollution in our oceans. Environmental Investigation Agency. Found at: https://eia-international.org/the-shocking-impacts-of-plastic-pollution-in-our-oceans (accessed June 29, 2018).
26. Harvey, Fiona. (September 15, 2015). Tuna and mackerel populations suffer catastrophic 74 percent decline, research shows. *The Guardian*. Found at: https://www.theguardian.com/environment/2015/sep/15/tuna-and-mackerel-populations-suffer-catastrophic-74-decline-research-shows (accessed June 29, 2018).
27. Chuck, Elizabeth. (January 28, 2016). Flint's next issue: What to do with empty water bottles? *NBC News*. Found at: https://www.nbcnews.com/storyline/flint-water-crisis/flint-s-next-issue-what-do-empty-water-bottles-n505781 (accessed June 29, 2018).
28. Baptiste, Nathalie. (September 28, 2017). Puerto Rico's drinking-water crisis isn't going away anytime soon. *Mother Jones*. Found at: https://www.motherjones.com/environment/2017/09/puerto-ricos-drinking-water-crisis-isnt-going-away-anytime-soon/ (accessed June 29, 2018).
29. Frostenson, Sarah. (March 22, 2018). America has a water crisis no one is talking about. *Vox*. Found at: https://www.vox.com/science-and-health/2017/5/9/15183330/america-water-crisis-affordability-millions (accessed June 29, 2018).
30. Malter, Jordan. (January 10, 2016). *CNN Money*. Found at: http://money.cnn.com/2016/01/10/news/economy/raw-ingredients-food-imports-safety-seafood/index.html (accessed June 29, 2018).

31. USDA Foreign Agricultural Service. (December 30, 2017). 2017 China's fishery annual. *Global Agricultural Information Network Report*. Found at: https://gain.fas.usda.gov/Recent%20GAIN%20Publications/2017%20China's%20Fishery%20Annual%20_Beijing_China%20-%20Peoples%20Republic%20of_2-17-2018.pdf (accessed June 29, 2018).
32. Siegner, Cathy. (April 24, 2015). Consumer Reports: Tests find 60 percent of frozen shrimp contaminated with bacteria. *Food Safety News*. Found at: http://www.foodsafetynews.com/2015/04/consumer-reports-tests-find-60-percent-of-frozen-shrimp-contaminated-with-bacteria/#.Wiq8LlWnHIU (accessed June 29, 2018).

CHAPTER FOUR

2030: A Plastic Tipping Point (Peak Plastic)

Patterns to Explain the Unexplainable

According to the late, great comedian George Carlin, humans are not only incapable of saving the planet; we are not even able to take care of ourselves! The planet is not going anywhere, Carlin tells us, but *we* are. Of course, the job of a comedian is to make people laugh, not in being a science expert; but consider that one of the most respected biologists of our era, Lynn Margulis, was in general agreement with Carlin's tongue-in-cheek assessment. Humans are simply just another species, Margulis told us, no different from runaway populations of bacteria, locusts, roaches, mice, and wildflowers that collapse and pose a threat to themselves but not the planet.[1] According to this perspective, humans have been programmed as just another life-form on a 4.5 billion-year-old planet where over 99 percent of all life that has ever lived no longer lives; life comes and life goes, impacted through this ancient ritual. Plastic might be one of our own inventions that helps to do us in, but it won't do much to affect the planet nor impact more than half of the earth's biodiversity that is simple bacteria. Margulis called this our "culture of delusion" when we believe we can control more than we actually can, including such actions as the use of a material we know can be threatening to us and the planet. Maybe if she were alive today, Margulis would scoff at Buffington's book *Peak Plastic* as yet another ploy or fit of arrogance that we can escape our path of evolutionary fitness that is leading to growth that will ultimately lead us to running off the edge of the cliff. According to biologists, the sixth mass extinction on Earth is currently underway, a "biological annihilation" due to human overpopulation and overconsumption.[2] Enjoy the ride!

Before you lose hope, or more importantly stop reading this book, recognize there is another point of view that finds us to be capable of free will with a large degree of control over our own domain. Norman Borlaug, an agronomist by trade, became a humanitarian who was largely responsible for the Green Revolution, the use of technological innovation to increase agriculture yields in order to feed a growing world population. In contrast to a fatalist view of man's extinction as a result of biology or our own ignorance, Borlaug considered the challenge to be an opportunity more so than an insurmountable problem. His successful initiatives led to an increase in crop yields in the developed and developing worlds through such technological innovations as the use of monoculture, which was met with great disdain by environmentalists who criticized these innovations as behemoth, corporate-type solutions that take away from culture and the rights of the community. To these criticisms, Borlaug bristled that if these environmentalists "lived just one month amid the misery of the developing world, as I have for 50 years, they'd be crying out for the tractors and fertilizers and irrigation canals and be outraged that the fashionable elitists back home were trying to deny them these things."[3] This is the voice of the innovator, the one who believes problems are challenges to be solved rather than an inevitable zero-sum game where we have to lose to win and vice versa. If Peak Plastic is a prescient warning of bad things to come, and I hope to have made this point in the first three chapters, how should we face this threat? As an inevitability; as our future as a function of some biological pattern that all species must face; as a zero-sum, fixed calculation that cannot be revised; or as a challenge to utilize technology and supply chain solutions? I think the answer to this question is obvious to most of us, hardcore biologists and stand-up funny men notwithstanding.

Innovation is the answer, the only answer. Why wouldn't a biologist believe in the magic of nature's supply chain through evolution, through innovation, in achieving growth rather than loss? The view from famous biologists, such as Richard Dawkins, is that humans are merely "lumbering robots" and not the by-product of billions of years of innovation that has led to a conscious being with the potentiality to program our society and planet in some ways. In how we view our future role in the world, I think this is more than just a philosophical or academic quibble because it brings to light whether or not humans are up to the task to keep growing in population and societal sophistication without ultimately destroying the ecosystem that is important for our subsidence. The negative impacts from fossil fuels, plastics, and other synthetic chemicals are not an endgame but a progression for us as a species. Perhaps if anything, we misunderstood the rules of how we should grow within the planet rather than being an example of how we are just the next species preprogrammed to destroy ourselves through our own evolution.

The mythical image of innovation is that it is magic, the solution that happens right before it is too late, such as the crafty prince finding a way to save the damsel in distress from the castle moat without the fire breathing dragon knowing, or the girl saved just in the nick of time by a dashing hero from the dastardly villain who tied her on the railroad tracks. This trope of the hero figure plays well also in the history of the innovators; the young Sir Isaac Newton sitting underneath an apple tree and getting hit on the head in order to discover gravity. Yet the story of what really happens is less exciting and is a story of countless failures before the innovator succeeds as an example of persistence and dedication. Thomas Edison is often mythicized as a magical genius inventor, but one of his most famous quotes was "98 percent of genius is hard work," which demonstrates how innovation really comes to fruition. Beyond the folklore and storytelling, innovation is almost always the process of a long and arduous road of experimentation and discovery that happens through teamwork, persistence, and often dumb luck. So rather than assuming our reckless path of economic growth and development is a function of our predetermination as a species, as a lumbering robot, it just sets us up to solve this problem through the use of innovation, the same tool that has been used to pull hundreds of millions and billions out of poverty, for our era to be the most prosperous and peaceful in history. But first, we must understand and acknowledge that this problem exists.

Innovation is a messy, flawed, and meticulous process that typically requires dumb luck of some sort, and often ends in failure. After World War II, mathematician Edward Lorenz chose to study meteorology at MIT after working for the U.S. Army Air Corps, and as the story goes, he accidentally discovered a phenomenon that at first seemed to be a simple mistake. As he was rerunning a weather pattern simulation on a computer system, he mistakenly ran it from the midpoint of the program rather than the start. As a result of a minor change in the value input of around 0.1 percent, a seemingly statistically insignificant value change, the results between the two models were dramatically different. Lorenz is not written in the annals of history as a divine genius, but as a result of understanding the mistakes that were made, he literally stumbled upon a grand discovery. He subsequently tested his equipment and found nothing wrong; after much experimentation, he concluded weather is not as perfectly predictable as thought, and moreover, small changes to parameters can lead to a dramatic change in results. These findings led to what is known today as chaos theory, or the butterfly effect, the proverbial flapping of a butterfly's wings in the Amazon jungle that eventually sets off a tornado in the United States. This field of deterministic chaos has become invaluable in the field of environmental science in attempts to explain the unexplainable, and it has also become important in helping us understand human behavior, a field not so easily understood through linear patterns. Not only is this finding of chaos theory potentially helpful in understanding how the plastic supply chain has evolved into such a difficult predicament of being so good and so bad at the same time, it is a model for how innovation can change the metaphor in understanding the patterns of behavior from butterflies flapping around the world (e.g., how plastic waste in Asia can harm us in the United States). Innovation is messy and a flawed process, but it can lead to *design for good* if we pursue it as such.

Another trait of innovation is it often seems to be not worth it, given a low probability for success at the onset, leading many to give up after the first try. Thomas Edison acknowledged he failed more than a thousand times, and Elon Musk is known for his failures as well, yet both these innovators are better known for their successes, even though they see it differently. Sometimes failure is a matter of timing; according to researchers, the success rate for innovative new products and services is likely between 10–20 percent, which is why we are affixed to miracles or doom more often than processes.[4] But what if our technologies and processes improved so significantly that we are able to detect patterns both from a myriad of data points never before able to be aggregated and from seeing the seemingly invisible in order to better understand our world? I think there's a lot of hope for the future in innovations to better balance the economy and environment, and it's not going to be abundantly obvious to most that this is worth tasking; even with better tools, we need to change the metaphor.

Science itself could also use an innovation makeover. The infamous science pariah Rupert Sheldrake may have some theories that border on pseudoscience, to state it kindly, but he does make a case that science is growingly dogmatic, becoming more like religion in that it can often encourage groupthink, leading to the banishment of those who challenge its conventional wisdom.[5] In matters of environmental policy around climate change and plastic waste, "scientific simplicity" can be dangerous in relying on a general theme that is logical, and maybe even provable to some extent, and filling in the details through faith and the greater good of saving the planet. In 1989, Noel Brown, director of the UN Environment Programme, warned that if global warming was not reversed by the year 2000, entire nations would be wiped off the planet.[6] Similarly, in 2006, the ABC documentary *Earth 2100*, hosted by Bob Woodruff, proposed that agriculture production would be decreasing by 2015 (it hasn't), and if greenhouse gases are not reduced by 2015 (they didn't), increasing global catastrophes will start happening in 2020.[7] Scientists who are too focused solely on the technical elements of a problem can paper over some of the details in hopes of the greater good rather than a model based on innovation that addresses an overall system cure of the challenges. A conventional recycling program and a technology solution that sweeps the plastic from the ocean are examples of piecemeal strategies endorsed by science that can prevent more system-based innovations from solving, not mitigating, the problem. Scientific simplicity and dogma must be avoided at all costs because they are life suckers to true innovation that is needed to solve challenges of this magnitude.

The real beauty in innovation, just as is the case with evolution, is in the details, those seemingly minor events that trigger growth and change. Thomas Edison did have over 1,000 patents, but it wasn't him alone who invented any of these ideas; he was perhaps the last contributor, tinkering on an idea built from many men until it became an invention and then an innovation. Edward Lorenz tinkered with these measuring devices and Norman Borlaug created 6,000 cross-developed wheat strains until he was able to assist in the crop losses in Mexico. This beauty of the role of innovation is the same of what seems so patently simple in biology and how changes occur, or do not occur; it is built upon a relentless trial-and-error method, seeking patterns that can hopefully soon lead to the cure of Peak Plastic with better tools. These intricate details in the patterns of human, economic, and supply chain behavior and activities over the past century have been neglected or unmeasurable, leading to incongruences in the overall plastic supply chain system of what has been good in the growth of plastic for our use and its downfalls, which are being addressed in this book. To solve this problem of Peak Plastic, the overall system of plastic must be analyzed and corrected in its detail, and then this "planetary crisis" is no longer an insurmountable problem.

An example from the past of how a public health problem wasn't handled as an overall system issue was the Cigarette epidemic of the 20th century and today. Prior to widespread cigarette smoking, the prevalence of lung cancer outside of industrial exposure was a rarity in humans, leading to an obvious association between the practice of cigarette smoking and an increase in the disease. As lung cancer rates began to rise in the 1940s and 1950s, scientific evidence began to emerge linking smoking and lung cancer, yet nothing happened of any substantiation. Cigarette manufacturers through their supply chain system urged for caution in judgment to the public as early as the 1920s when science began mounting these health concerns. Advertisements were focusing on claims its products were healthier than those of the competition because more doctors used their brand, or their product led to fewer irritated throats.[8] Through such campaigns, it didn't appear that cigarette manufacturers were challenging that these products were not possible to cause health-related concerns, but rather that there was no direct causal link between one activity and a specific health outcome like lung cancer. Cigarette manufacturers and other parties in the supply chain would often invoke the perspective of "plausible deniability," suggesting they either were not aware of any health concerns or could not validate or invalidate every possible outcome of a consumer's use of the product and that person's circumstances. For example, two different individuals who smoked for decades might have different outcomes; one may live to be 90 while the other dies at 60 as a result of lung cancer. Even now, after decades of evidence tying lung cancer death to smoking, there has yet to be a study that causally proves smoking to lead to lung cancer even though it is well accepted by almost everyone within the culture of the correlation.[9] This is a prominent example of how scientific research and public policy outcomes can be a slippery slope in today's modern society and industrial supply chain system. Another example that is relevant today is the role of the supply chain system and public health policy related to America's opioid crisis, a menace that is leading to 115 deaths a day. Who is responsible for this, the drug manufacturer who makes certain claims about the product safety, the distributors who use cushy tactics to get doctors to choose their medications, retail operations like pharmacies with lackluster control systems, or the users themselves who become addicted?

In too many cases, the American public has expected a government agency of some sort to protect them from the health issues caused by products like cigarettes, opioids, and now maybe plastics? By 1964, decades after smoking was known to be a health problem and decades prior to the problem remaining today, the surgeon general of the U.S. government mandated for labels to be placed on cigarette packaging that identified smoking to be dangerous to your health. Today, the fallout from this passive approach is devastating and deafening: according to the U.S. Treasury Department in a study conducted in 1999, there are 45 million Americans who smoke and approximately 400,000 deaths annually, placing a high cost as well on the healthcare system of $45 billion annually, probably a much higher number today in accounting for higher percentages to be treated and inflation.[10] Worldwide, there are 6 million cigarette-related deaths a year, on track to reach almost 8 million by 2030, with costs to the global economy of $1 trillion, much higher than the $269 billion collected in tobacco taxes, according to the World Health Organization (WHO).[11] The costs in life and treasure to the American public of these epidemics have and continue to be staggering. In the United States, public service campaigns that brought attention to the problems with cigarette smoking as early as the 1970s has done nothing to deter an increase in teen smoking rates since 1988, a statistic noted in Malcolm Gladwell's book *The Tipping Point: How Little Things Can Make a Big Difference*. Simply telling a teen not to smoke is as effective as reminding us all to recycle our plastic—a blind adherence to a policy without an understanding of it—and the absence of social, cultural, economic and supply chain is the lesson of how hell is paved with good intentions. Similarly, warning the public against the problems of painkillers has not prevented the United States from consuming 80 percent of the world's supply, another example of a supply chain system without proper controls.[12] This opioid epidemic facing the United States has consistent patterns in its supply chain system as did (does) the cigarette and, I would make the case, the plastics industries. There is evidence that there was at one point a sufficient volume of opioid prescription drugs in the United States for a three-week supply for every man, woman, and child in our nation.[13] Similarly, every American throws away more plastic (185 pounds) than our average weight, another example of how supply drives behavior.[14] Whether cigarettes, pain medication, or even plastic, the moral of the story is while we need the scientific community and the government more than ever to be a cornerstone in addressing these great problems, it has not been, and will not be, enough to truly create disruptive innovation to the model. If the problem is identified by the scientific community as an inherent danger and yet the supply chain system encourages demand as a function of a large supply, the problem will persist. The solution to these challenges is not more faith in science or government but rather more innovation to gain a deeper empirical knowledge.

The Possibility of "Doing Nothing"

Put simply, public policy strategies can be effective in educating the public if they wish to, but there is no clear causal evidence of how any sort of jurisdictional control will solve any broad public environmental, health, and safety challenge such as cigarette smoking, painkiller abuses, climate change, gun use, and the overuse of plastic and how it is affecting us and the environment. Recycling advocates in the United States have always promoted a government-mandated approach akin to a Swedish-style approach to plastic waste that is too expensive to be viable in a developing nation, and even in many developed nations where recycling rates are in the single or low double-digit range while plastic production and use grows at a 9 percent rate every year. So what do we do, nothing? It's often the response back to me from these recycling enthusiasts when I debunk the economics of these bottle bill programs. Unfortunately, for most Americans, *doing nothing* more than voluntary recycling programs is the most popular approach to this problem that many do not understand. It's similar to other problems not well understood; for example, while the United States now produces less air pollution domestically, it has experienced a 65 percent increase in ozone concentration as a function of problems in Asia, a problem many believe we have no control over.[15] Heightened ozone levels in the United States reaching both the East and West coasts and stretching as far inland as Colorado will have an adverse impact to our health, yet what can statewide or regional public policy efforts accomplish by raising these concerns across a global supply chain model? The water pollution crisis facing China and the rest of Asia impacts us as well, whether we know this to be true or not; it has been estimated 80 percent of China's water from underground wells is polluted, which is the same water utilized for farming and industry across its heavily populated plains, a significant element of the global supply chain.[16] This polluted water is recycled back to us as consumers in the global supply chain and through the runoff into the oceans, including the solid waste pollutants of plastic. The lack of regulation of the imported seafood coming from Asia via contaminated aquafarms, plastic products and runoff into the oceans should make us think twice of the ramifications of the global supply chain and our need for greater innovation given the human path of population and consumption growth. But what can really be done by our governments about such problems in a global supply chain model? Is it a lost cause?

For better or worse, public policy initiatives and corresponding agencies and programs are probably not adequately protecting our environment, health, and safety, and the only countervailing approach is to do so by impacting economic growth, an unpopular stance in both developing and developed economies. Consumer-citizens are given the difficult options of sending millions of tons of plastic every year to landfills, seeping into groundwater, releasing chemicals inadvertently into our rivers and lakes, or crippling regulations to economic growth and costs. Likewise, consumer-citizens cannot and should not rely on our supply chains to protect them as well; the routine release of micro-plastic residue from our dishwashers and washing machines either in the form of cleaners or leeching from the articles being cleaned that has not been conclusively proven one way or the other to be safe or toxic to our water systems. If 60 percent of cancer deaths are caused by smoking and diet, what is responsible for the balance of the remaining 40 percent? According to a 2010 report issued by the President's Cancer Panel, the decades-old estimate of 6 percent being due to environmental and occupational exposures, such as plastic, is outdated and far too low.[17] Cancer rates are increasing in the developed world, but it doesn't appear to be clear as to why, especially when lung cancer rates are falling.[18] Who is responsible for addressing this mystery, or is nobody responsible because the questions are not being asked? Of the National Institute of Health's list of 54 known human carcinogens, each of them cause at least one type of cancer that may not be well understood at this time and is likely underestimated.[19] These are the public policy questions that appear to be compounding, not just within our own nation but impacting us as a one-world planet, growing in concern as our use of synthetic materials within these global supply chains grow as well. Can we *do nothing* if public policy is our only option, or can we unleash the innovators to solve these multidimensional problems tying environmental sustainability with economic growth? Can Peak Plastic be addressed through a model of innovation?

The Message. . .to Who?

What is wrong with our scientific research organizations, public policy, supply chain systems, and innovation engines when outrage and public policy interventions are only called for after there is a delay in addressing the problem? Cigarette smoking and lung cancer, the opioid epidemic, and the thousands of related deaths every year are proof that a new approach is required, a private-based approach of innovation as our only hope. But rather than a complex and slow-moving approach to innovation, the model needs to be corrected to a *pull* rather than a *push* model to incent the entrepreneur instead of prompt the regulator. Sure, grassroots efforts can lead to finite wins, such as the banning of Styrofoam cups in some localities or an EU target to ban single-use plastics by 2030, but these are merely drops in the bucket related to the overall problem, otherwise known as a distraction. If the scientific method is how we should ask and study questions, then the clear evidence from the past demonstrates the need for a new approach, an approach directed toward the innovators and not government agencies, organizations, and the general public. In regard to this plastic problem, it is not going to bubble up to the top of the public policy initiatives until it is too late; therefore, instead of a process that is only jumpstarted from outrage, with money thrown to solve it in a reactive mode, induce the innovators through proactive models with clear cost benefit frameworks in the supply chain markets. The focus needs to be directed to the innovator as the problem solver.

The focus also needs to be on the impassioned environmentalist, not as a problem solver but rather as a message deliverer. Rather than to push for government legislation through guilt trips appeals without viable cost-benefit models, invite them into a new solution paradigm of the innovation engine to solve rather than mitigate problems. These protagonists can also help educate the markets as to how science, innovation, and supply chain can drive sustainability rather than marketing and regulation advocacy. Take for instance Michael Braungart, a German scientist who does not fit the mold of a European environmentalist type because he advocates for a manner of how the design for good must happen in product development, all the way through the supply chain, in order to improve the economy and the environment. Dr. Braungart is a new type of environmentalist who is focused on the life cycle as instrumental to the supply chain and vice versa. His message appeals to the entrepreneur and the innovator, not the bureaucrat and the policy maker. Rather than appealing to governments or its citizens, our message of Peak Plastic and its challenges must eventually get to the innovator and enable them to be incented, and this drive can only come from those who are most passionate and knowledgeable of the topic. This is an *inside job*.

The general public has often been the target of plastic use abuse, yet despite such guilt trips, these consumers will only change their behavior as a result of innovation improvements without losing anything in return. In reality, Americans are growing increasingly tired, bored, or both with the environmental movement: a 2016 Gallup poll found that only 42 percent of Americans considered themselves to be an environmentalist, a label that has been on a declining path since 1991 when it was at 78 percent.[20] According to this poll, the primary reason for the decline of environmentalism is as a political association, meaning it has become more of an ideology than a practice to provide solutions in improving the environment as a result. It's easier for an environmental advocate who is a celebrity to preach for the care of the planet without an understanding of its economic impact on the ordinary citizen who cannot afford such luxuries. This is why the basis of this book is for supply chain systems to address the problems of Peak Plastic, not as an ideology, but rather as a model of practical solutions to improve both the environment and the economy. These waste challenges won't be solved by consuming less and recycling more, but rather, the problem must be resolved through the innovation engine.

Therefore, you are not expected to make the dismal choice of choosing between the biologist's perspective of accepting yourself as a *lumbering robot,* leading our species to destruction, and the complete elimination of evolution in banning plastic to prove that we are otherwise—it will be science, technology, innovation, and supply chains that will fix what's wrong with plastic without impacting its good. A holistic approach to a supply chain that was not in place at plastic's onset will provide a more balanced use between the environment and economy; Peak Plastic is a supply chain problem and not a material science or recycling problem per se. Innovators and supply chain professionals will drive the change in a manner material scientists and policy analysts cannot; it is not the invention but rather the innovation of the plastic bottle that is the challenge, not the single bottle but rather the 499,999,999,999 other bottles this year in the world that were produced by the system without an adequate plan of what to do with them, not to mention the long rows of other cheap plastic products driven by Walmart's and others' global supply chain system. Solving these challenges, especially in the fast-growing megacities and slums in the developing world, such as Manila, Jakarta, Mexico City, will require urgent and practical approaches. Likewise, even the problems beginning to sprout in America's inner cities can be solved by supply chain solutions, as will be addressed in the next chapters of the book. It becomes a matter of linking our industrial supply chains to the supply chains of nature in order to find innovations that profit both rather than choosing one over the other. It's a paradigm shift in thinking of how to solve challenges, and that's our message to center on related to Peak Plastic, hoping to spread the word in two years in order to implement in 10 years.

Yet keep it in context; it is not a matter of the consumer not participating in recycling programs, governments not enacting common-sense legislation, or material scientists not seeking new designs, but rather it is the message of who should be driving the bus, and that is absolutely the same bus driver who brought the products to town. For the past fifteen years of my career, I have been a supply chain leader for a large U.S. company, working toward distributing manufactured products in this linear fashion that can be described as a lack of a closed-loop solution. Once the goods reached the consumer, it was up to the waste management and recycling companies to take care of the rest. About 10 years ago, I took a sabbatical to Sweden to understand how to fix this problem of waste. Sweden, of course, is known as the darling of the environmental crowd, the nation of zero waste and obsessive conformance to recycling. It was my expectation I would finish my PhD and replicate this system and culture to the United States, but what I learned was different from what I thought I would learn. Much like Lorenz, I had a "gee, that's odd" moment when I realized Sweden's approach was simply well-executed mitigation that only slows down the environmental and economic waste problem. As I thought through and experimented with various options, it became clear to me that waste management is simply an after-the-fact Band-Aid addressing the inefficiencies of the overall supply chain solution. Since then, I have been developing solutions that will be discussed next in this book and demonstrate how supply chain systems can solve the problems. At least for me, and those whom I have collaborated with on my research as well as industry partners, the message is clear: we are trying to fix environmental challenges in the wrong manner, and what we need to do is to change our approach to get on the right path.

Therefore, this message of Peak Plastic is not a callout of shaming to consumers or even to the current manufacturers of so much unnecessary and damaging packaging waste. They are simply the drivers of the current-state supply chain required to keep the economy growing. It is a callout to those who can design products and supply chain systems for good and connected to nature, which will lead to new markets and opportunities. In the future, it will be big business to *design for good*, and as a result, innovators will be attracted to developing these new markets. The goal of this book is to make this happen before it is too late.

Calling All Supply Chain Innovators (2020–2030)

It's official: consumers, you are off the hook; you should continue (or start) to participate in recycling programs to eliminate plastic waste in the environment, but you will no longer be considered the culprit responsible for this mess and made to feel guilty for your usage. You purchase disposable materials that cannot be reused, and you are simply the benefactor of these wondrous supply chains that are capable of providing these products to you at such low prices and high quality. Nor can you do anything about it other than recycle (mitigate), and our goal for you in the future is to educate you to what is happening rather than to guilt trip you in broad, incomplete messages. The focus will be on the innovator, those who have the role in our economy, and society for that matter, to drive the change. Groups proposing bans, bottle bills, and recycling as solutions should shift their focus from mitigation schemes to advocate for innovation instead, to *design for good*, and supply chain solutions rather than legislation. Governments and even intergovernmental organizations must spend less time making proclamations and mandates and use their resources, mainly financial, to incent innovators to fixing these grand challenges. Even innovators have to get better: nothing against the creative efforts of entrepreneurs such as The Ocean Cleanup project, but funding should follow science and supply chain and be funneled toward source reduction strategies rather than mitigating waste after it has floated away for thousands of miles. According to Boyan Slat, the young Dutch innovator who dreamed and implemented this project, the focus is on cleaning the five gyres of the ocean where much of the plastic waste eventually heads to, therefore "letting the plastic come to us." According to Marcus Eriksen, the well-known expert who took the issue of plastic ocean waste and made it a public news event, this approach is a distraction away from the upstream aspects of the supply chain that need to be addressed.[21] Eriksen is certainly correct: while Slat's project is cause célèbre, it has taken focus away from projects designed to prevent plastic from entering the ocean in the first place. Again, the focus needs to be on innovation incenting innovators in the right manner to solve this plastic problem over the next 10 years.

Why not incent supply chain innovators to solve these problems? Rather than seeking scientists to design new plastic bottles or plastic polymers, how about a call to all supply chain innovators, the next Jeff Bezos (or the current one), to define how the supply chain system looks and then design the products? Inventors of technologies are important, such as Boyan Slat, but these contraptions to collect plastic waste in the middle of the ocean must become an aspect of an overall holistic solution rather than just one means to one end. Technology by itself is not going to fix these behavioral and systemic problems found throughout the supply chain, including how a product is designed, manufactured, distributed, used, and post-used. In contrast, a supply chain innovator will seek to understand the incentives and behaviors of all agents in the supply chain to understand the root-cause problems, likely having little to do with the parameters established by Slat to design his invention. After all, it was not the designer of the plastic bottle who created the problem but rather the supply chain systems in place to enable its massive proliferation across the globe; it was not Nathaniel Wyeth, the DuPont scientist who invented the bottle, but the superefficient supply chain operations of Nestlé, Coke, and Pepsi who are both the innovators and culprits. It is these brilliant minds of an end-to-end supply chain solution—the Jeff Bezoses, the Sam Waltons, the Jack Mas, the greatest creative minds—who can build the model of the future of plastic, an opportunity of all of our lifetimes!

These supply chain innovators will likely not be from the same mega supply chains that produce, ship, and distribute billions of plastic units a day; these companies are entirely focused on what the consumers are demanding—more products at lower prices and higher quality and selection. I have spent a great deal of time in my career as a leader for a private practice supply chain and researcher to understand that these successful manufacturers, suppliers, service and solution providers, and retailers are laser-focused on what their consumers want in the current state. These companies are undoubtedly concerned about the environment, despite what their detractors might say; they simply have to be really focused on improvements in order to maintain their sales, market share, and profitability. Today, the wrong policies are being promoted due to a lack of understanding of the problem, which is the plastic supply chain system itself. Through a grassroots campaign, change programs should be implemented through the same advocacy groups already in place, but instead of promoting public policy change, they should be seeking the supply chain innovators. During the book tour for my last book, *The Recycling Myth*, I found most of these advocacy groups are well suited to drive change if it is appropriately directed; partnering with these groups as well academic and business innovators will build the groundswell of support required to drive the change. Also included in the mix will be stakeholders who are already being impacted by Peak Plastic well beyond the 2030 date; this is mostly the poor and marginalized in the megacities of the developing world and some of the Rust Belt cities in the United States who have never recovered from the rapid decline of manufacturing in the 1970s and 1980s. I have already been working with some of these groups to create the change, and this amalgamation of stakeholders builds the grassroots plan for how change can occur within and to the supply chain system. Today's supply chain system is the problem due to how it is designed: the linear, one-way model doesn't take into account much other than getting the product into the consumer's hands as quickly, cheaply, and most efficiently as possible. The new supply chain system will redefine what it means to be cheap or not, what the experience is, and to whom it reaches.

The 2020 decade has all the makings of being one of the most transformational periods in human history: the transition from carbon-based to renewable energy, the potential to transform how we grow and raise food sustainably to protect the planet as well as to feed a growing population, and a greater understanding of our own human bodies and all materials through nanoscience. Following this same leap forward, it also appears to be an excellent decade to erase the contradiction of plastic in our lives in a manner that is transformational as well. One decade is a rough-cut plan driven more by the timing of the urgent need for action than it possibly might be regarding what can get completed in short order. In this book, I will discuss five solutions as a starting point to address these challenges from plastic, starting with a classification system for an improved understanding for producers and consumers and then four types of innovation: design, technology, supply chain, and nanoscience. From these, we can build the change necessary to turn the tide without requiring us to change how we live our lives in our synthetic world.

2030: One Supply Chain/Ecosystem

In 1909, Nikola Tesla predicted one day it would be possible to transmit wireless messages over the world so cheaply that any individual could do so on his own apparatus. Tesla's vision would one day come true, but it would take nearly a century for it to come to fruition. In contrast, John Kennedy gave a speech in 1961 about how America should commit itself to landing a man on the moon by the end of the decade, a prediction that was not only a vision but a course toward action that was fulfilled during the stated time period. Is a prediction made by a UN director named Noel Brown in 1989 regarding the ability of climate change to wipe us off the planet by 2000 meant to inspire or scare us? Kennedy's message inspired the American people to action while Noel Brown's failed, thankfully, and did nothing to move us in the right direction. Instead of the tired messages of gloom and doom, how about a message of hope to address Peak Plastic, in putting in place a supply chain system by the year 2030 that is one with our natural ecosystem? You bet. People (particularly Americans) love a good challenge, especially one that demonstrates exceptionalism and national pride. The same message of hope can work for any nation willing to take on their Peak Plastic challenge in the right manner.

With so many other magnificent transformations predicted to occur over the next decade or so, I believe there is an ample case to be made for why 2030 should be the right timeline for our supply chain systems to be one with nature, as an inspirational target calling out for all supply chain innovators to address. Peak Plastic is happening today in China, the Philippines, and other nations in the world; one could make a case that it is happening as well in U.S. cities like Flint, Michigan, and territories like Puerto Rico. It might not be the problem everyone is focused on, such as a lack of environmental regulation in Asia, hyper growth in megacities, a failed municipal water system, or an epic hurricane event, but it's there if you look for it, and it will (unfortunately) be easier to detect as time goes on. If we fall into the same traps as we've fallen into in the past, we often only get clarity in our vision when it is too late, and in some of these circumstances, it's too late. So let's not wait for things to get much worse, or spread out further in dereliction.

Not only is this chapter a paradoxical story about how the advancement of our societies has led to enormous economic growth and prosperity in the last few decades at the risk of environmental catastrophe, perhaps unnoticed, it is also a segue between a current state of plastic good, plastic bad, and the future of plastic as in the rise or fall of our synthetic world. The first three chapters are relatively simple: identify the lingering questions regarding our use of plastic in order to form a rallying cry that something must be done. But what should be done and by whom? The answer to this question is clear: provide a clear vision of what needs to happen and by when through these agencies who advocate to eliminate the problems and incentivize the supply chain innovators to figure it out. With a targeted date of 2030, encourage our supply chains, scientists, and innovators to solve the problems rather than doomsayers portending policies of failures past. Peak Plastic needs to be a call to action to transform our supply chains and, in doing so, our ecosystems; if our oceans are connected as such, there will no longer be a need to send robots to these ocean gyres to collect plastic at the end of a 2,000-mile journey. There will also not be a need to debate what is more important, our economy or the environment, because they will be linked as one. But for this to happen, we must think exponentially rather than linearly. As what happened to Edward Lorenz in his weather computer simulation lab will hopefully happen to us: rather than considering these plastic articles as statistically insignificant as an individual article, we begin to understand how trillions of them amid our vast planet or nearly 25,000 miles and over 7 billion inhabitants, not even counting the trillions of non-human beings, are being transformed to this significant event worthy of our attention called Peak Plastic.

References

1. Margulis, Lynn. (September 1, 1998). Life on Earth doesn't need us. *The Independent*. Found at: http://www.independent.co.uk/arts-entertainment/life-on-earth-doesnt-need-us-1195424.html (accessed June 29, 2018).
2. Carrington, Damian. (July 10, 2017). Earth's sixth mass extinction event under way, scientists warn. *The Guardian*. Found at: https://www.theguardian.com/environment/2017/jul/10/earths-sixth-mass-extinction-event-already-underway-scientists-warn (accessed June 29, 2018).
3. Tierney, John. (May 19, 2008). Greens and hunger. *New York Times*. Found at: https://tierneylab.blogs.nytimes.com/2008/05/19/greens-and-hunger/?pagemode=print (accessed June 29, 2018).
4. Fisher, Anne. (October 7, 2014). Why most innovations are great big failures. *Fortune*. Found at: http://fortune.com/2014/10/07/innovation-failure/ (accessed June 29, 2018).
5. Sheldrake, Rupert. (2013). The science delusion: Freeing the spirit of enquiry. London: Coronet.
6. Morano, Marc. (September 20, 2016). Failed tipping points: History keeps proving prophets of eco-apocalypse wrong. *Climate Depot*. Found at: http://

www.climatedepot.com/2016/09/21/history-keeps-proving-prophets-of-eco-apocalypse-wrong/ (accessed June 29, 2018).

7. Danner, Alexa. (May 29, 2009). "Earth 2100": The final century of civilization? *ABC News*. Found at: https://abcnews.go.com/Technology/Earth2100/story?id=7697237&page=1 (accessed June 29, 2018).
8. Lawrence, Leah. (March 10, 2009). Cigarettes were once "physician" tested, approved. *Healio*. Found at: https://www.healio.com/hematology-oncology/news/print/hemonc-today/%7B241d62a7-fe6e-4c5b-9fed-a33cc6e4bd7c%7D/cigarettes-were-once-physician-tested-approved (accessed June 29, 2018).
9. Proctor, Robert N. (2012). The history of the discovery of the cigarette-lung cancer link: Evidentiary traditions, corporate denial, global toll. *Tobacco Control* 21 (2): 87–91.
10. U.S. Department of the Treasury. (1998). The Economic costs of smoking in the United States and the benefits of comprehensive tobacco legislation. Washington, DC.
11. Miles, Tom. (January 9, 2017). Smoking costs $1 trillion, soon to kill 8 million a year: WHO/NCI study. *Reuters*. Found at: https://www.reuters.com/article/us-health-tobacco-idUSKBN14T2H5 (accessed June 29, 2018).
12. Gusovsky, Dina. (April 27, 2016). Americans consume vast majority of the world's opioids. *CNBC*. Found at: https://www.cnbc.com/2016/04/27/americans-consume-almost-all-of-the-global-opioid-supply.html (accessed June 29, 2018).
13. Lopez, German. (December 21, 2017). The opioid epidemic, explained. *Vox*. Found at: https://www.vox.com/science-and-health/2017/8/3/16079772/opioid-epidemic-drug-overdoses (accessed June 29, 2018).
14. D'Alessandro, Nicole. (April 7, 2014). 22 facts about plastic pollution (and 10 things we can do about it). *EcoWatch*. Found at: https://www.ecowatch.com/22-facts-about-plastic-pollution-and-10-things-we-can-do-about-it-1881885971.html (accessed June 29, 2018).
15. Chappell, Bill. (March 3, 2017). Smog in western U.S. starts out as pollution in Asia, researchers say. *National Public Radio*. Found at: https://www.npr.org/sections/thetwo-way/2017/03/03/518323094/rise-in-smog-in-western-u-s-is-blamed-on-asias-air-pollution (accessed June 29, 2018).
16. Buckley, Chris, and Vanessa Piao. (April 11, 2016). Rural water, not city smog, may be China's pollution nightmare. *New York Times*. Found at: https://www.nytimes.com/2016/04/12/world/asia/china-underground-water-pollution.html (accessed June 29, 2018).
17. Israel, Brett. (May 21, 2010). How many cancers are caused by the environment? *Scientific American*. Found at: https://www.scientificamerican.com/article/how-many-cancers-are-caused-by-the-environment/ (accessed June 29, 2018).
18. Jones, Greg. (February 4, 2015). Why are cancer rates increasing? *Cancer Research UK*. Found at: http://scienceblog.cancerresearchuk.org/2015/02/04/why-are-cancer-rates-increasing/ (accessed June 29, 2018).

19. U.S. Department of Health and Human Services. (2016). 14th report on carcinogens. Found at: http://app.knovel.com/hotlink/toc/id:kpRC000063/14th-report-on (accessed June 29, 2018).
20. Jones, Jeffrey. (April 22, 2016). Americans' identification as "environmentalists" down to 42 percent. *Gallup*. Found at: http://news.gallup.com/poll/190916/americans-identification-environmentalists-down.aspx (accessed June 29, 2018).
21. Stokstad, Erik. (May 11, 2017). Critics say plan for drifting ocean trash collectors is unmoored. *Science Magazine*. Found at: http://www.sciencemag.org/news/2017/05/critics-say-plan-drifting-ocean-trash-collectors-unmoored (accessed June 29, 2018).

CHAPTER FIVE

The Alternative to Peak Plastic: Exponential Thinking

The Recycling Myth

After spending three years of my life shuttling between Colorado, where I worked and lived, and northern Sweden, where I was working on a PhD, I was excited about putting this research to work. Why do people recycle in Sweden, but not in the United States? I returned with more questions than answers, and that was not necessarily a bad thing because each question led to a new door opening and to new paths to consider. Close to the Arctic Circle where I went to school, I would study these socially responsible Swedes as they did their daily routines that included collecting, washing, sorting, and then returning these containers to a mandate redemption site for what reason? I would ask them myself, and it was clear that it was due to acculturation more than a solid economic business case; few if any really understood the technical or economic reasons behind what they were doing. The average Swede believed that these required tasks were being done for the good of the environment, but beyond this slogan, few that I spoke to really understand the details of what this really meant beyond the concept. In fairness, I think the Swedes were more focused on what would happened if they did not recycle rather than when they did, meaning the goal was to prevent littering and landfilling rather than a clear understanding of reuse. For the older generations, it became a daily routine as brushing their teeth and taking a shower, but the younger Swedes seemed to be more skeptical, leading them to be considered as selfish and too focused on consumption by their parents. I recall some explanations countered that it is sometimes about doing the right thing, even if you do not know why, an acknowledgment to doing what's best for the group rather than individualism that can be seen as selfishness, but can often lead to innovation.

Even in the United States, recycling can be a sacred cow, and why not? The stated alternative to recycling is throwing the article in the trash can rather than attempting more innovative solutions, which is the subject of this book. Blind cultural adherence leading to high recycling rates is seen as the best of possible outcomes, which is the real fallacy of the problem. Once I finished my studies in Sweden, I was burdened with this paradox, straight from the data, when I actually hoped instead to find a way to change the culture in the United States to match the Swedes! My first attempt at discussing these perplexing findings within the research community was as I was expected to do: to the peer-reviewed journals where concepts are to be vetted within the research community across the world. I do agree with the idea of this: peer-reviewed journals are intended to take on a level of professional scrutiny not available in other avenues, addressing significant issues using thorough methodologies among the most learned individuals in the world. Yet like any other group of like-minded individuals, it can become an echo chamber of protocol and hierarchy, with knowledge being built on the shoulders of giants sometimes favoring the dominant design over innovation. From being a part of this community, I was fully aware of the challenges: one, a tidal wave of cultural conformity exists in academia regarding recycling and sustainability, and two, the difficulty in publishing a paper with as wide of a path needed to address an overall system rather than a slice of an area of expertise was outside of the expected protocol of a researcher. From my experiences in Sweden, I knew the potential backlash in being considered anti-recycling, especially since I work in the consumer products industry, yet I was hoping a solid research methodology would lead to a consideration of a new path, based on the data. Not surprisingly, my initial submissions were failures right out of the gate as they were "desk rejected," meaning the editor did not believe the topic, method, and/or structure of the paper was worthy to be reviewed by reviewers of the journal. After asking a lot of questions of editors and modifying my research to not attempt to accomplish so much in a single paper, I got on track with publishing some papers, but with a much narrower scope and audience than I was hoping in order to turn the tide. Eventually, I came to believe this path was too slow moving of one that was required to create a paradigm shift in how the consumer products supply chain was interfacing with consumers and the environment, and I decided that I needed a different medium to frame this problem and its potential solutions. I focused on the nonfiction book market to start the story as to why recycling programs do not and cannot succeed.

My last book, *The Recycling Myth: Disruptive Innovation to Improve the Environment*, gave me the platform to ask the question to a broader nonbusiness audience: How many of us really understand how today's supply chain system was created and needs to solve this recycling problem? The question was a rhetorical one: the consumer not only has a low understanding of what is plastic (and what it is not), he or she also does not understand the supply chain system that is able to deliver it to us as it does. Because this is not understood, and it is clear from interviews and audiences that it is not, then how does anyone understand whether putting a used plastic article into a container is going to accomplish anything? To my surprise from my book tour, there was little emotion and controversy regarding the point of view that recycling is a myth, even in the recycling community! Of course, an advocate will advocate for it, and consumers are considered to be careless if they chose not to participate, but there was little controversy surrounding the ineffectiveness of the sacred cow of recycling despite this is what many of them are trying to convince the general public! The rationale behind this apparent paradox is that while recycling is known to be ineffective, it can be effective if more consumers participate and states legislate mandatory participation! To summarize: recycling is viewed as a sacred cow of environmentalism, and yet many experts in the community realize that it cannot succeed. Plastic production and use continues to grow as recycling rates remain stagnant. The environmental and recycling industries, as well as many researchers, are unsure of what next if not recycling programs, and the answer is a *call on all supply chain innovators*, as I am advocating for as a solution in this book. Based on my research and experiences over the past decade, there is no other way to look at this situation!

It's time for everyone to get past this obvious contradiction so we can move on to the next approach, that of supply chain innovation! The good news is this: sheer hopelessness embodied in these individuals as they face a problem that they cannot control will soon come to an end! During my publicity tour for *The Recycling Myth*, I would see the changing of emotions as I started my message describing the current state and feckless nature of recycling and the future path that required supply chain innovation; there was an uneasiness during my data dumping start of the message that clearly showed the hopelessness of recycling programs in their current state. Many people did not want to talk about it because it was so empirically logical from the data that recycling does not and cannot succeed, yet we see the damage that is happening as a result of runaway plastic. After making this point, I would pivot to the second message of the book of the need for disruptive innovation in order to solve the root-cause problem. This was normally the transition point with the audience of passive acknowledgment to smirking, and even some to ridicule; one prominent waste-management company executive told me in a nice way to "enjoy the walk back up to my ivory tower," a reference to ideas like this sounding better in a book than being plausible in real life. The last speech I officially gave for the book tour was to another group of professionals in the waste and recycling industry where I ended the talk with this rallying cry toward disruptive innovation to save the day; what exactly does that mean? After the polite round of golf claps, and most of the audience shuffling to their next session speaker, a few true recycling geeks lingered around to ask me detailed questions about how I believed these solutions would begin taking root in the next five years. One leader of the conference slapped me on the back said, "Thanks for the talk, and make sure you come back when you figure out what these big solutions are all about."

From this journey, I began to understand this plastic problem as more than just one of a material science or even supply chain journey. It would need to be solved beyond cultural shifts and public policy and also look at psychological foundations for consumers to adapt to what's happening today given these problems are larger than their span of control. Or would it? Is the problem still lurking inside these consumers, the same individuals who happily purchase these single-use synthetic products who had no part in designing, manufacturing, and fulfilling them? It seems like a constant losing battle to pin the burden on the consumer to carry the burden of saving the environment or in guilt trips for them to stop. Or even pinning the blame to our industrial systems that are able to bring these products to consumers in so much efficiency that it becomes wasteful. Blaming industry is too easy as well and becomes a fallacy. The journey of plastics and recycling should be less about who is responsible and more about whether we are on the right path in the first place; the focus must be on getting on the right track to solve, not mitigate, this problem!

Once we switch to the correct path as was discussed in the last chapter, the sometimes quiet and uninvited voices of innovation are out there to be heard. One notable event happened to me when a chemistry professor at a private university in Mexico City had his class read *The Recycling Myth* as part of his class. According to his recollection, the students in the class were so shocked to learn how these supply chain systems operating in the United States were no different from the inefficiencies that they saw in Mexico, and that these systems were actually designed in this manner! The students asked me to speak at their Sustainability Conference, and I did, to answer so many of their questions as to why this was happening, how it was happening, and what could be done about it. As I normally find, it's great to see how the next generation is able to frame up the problems differently from us elders to consider options of what has not succeeded in the past. After I spoke, the chairman of the conference, the same professor who asked his students to read the book, gave me a tour of a lab on campus where they chemically recycled plastic bottles into high-grade chemicals. As we walked to the lab, I was expecting to see some sort of academic concept that works in theory but not practice. What I saw instead was an innovative and effective technical design for how to recycle plastic bottles that was sitting there, on a university campus in Mexico City, unused. Yes, the university had discussions with a large beverage manufacturer before they politely declined given a lack of scalability of the technology. I tried to reconcile just exactly what I was seeing, and why this invention was sitting there, unused, rather than becoming an innovation. The technology seemed to work and was effective, but it could not work within the existing supply chain system. Was it the technology or the supply chain that was the problem then? Sometimes solutions can be the last place you'd expect, and the problems could be sitting right in front of you. More about that technology in the solutions section of the book.

The moral of the story is this: if you clear the air, listen, and solicit for change, the innovators will arrive to work on the problem. If we are under false pretenses that recycling programs or technology devices by themselves will solve for it, then we are distracting the innovation engine from doing its work. Currently, it is either the wrong people involved or focused on the wrong solutions that is distracting the right people and solutions away from addressing them. So let's clear the playing field, focus everyone on the same problem to address, and then incentivize the innovators to make the solutions happen.

David and the Two Goliaths

Before we get too far down the path of how innovation is going to save the day, it's worthwhile to discuss the modern-day consumer's love-hate relationship with Big Oil; consumers can rant all they like about the huge influence and geopolitical crises associated with our dependency on petroleum, but we are the ones as consumers who are making these choices. In 2016, industry revenues totaled $1.7 trillion, or almost three times more than the markets of the major metals combined, which equals $660 billion annually.[1] Evidence of this love-hate dynamic is as follows: worldwide demand is expected to increase by 30 percent by 2035, and there is an expectation to offset this through greater gains in fuel efficiency for years to come, leaving the industry in a bit of a quandary.[2] Today, over 80 percent of the world's energy requirements are met by fossil fuels, but this slice of the pie will shrink through the use of renewable energies and disruptive designs in battery storage.[3] With traditional crude prices falling through gains in natural gas (1.6 percent growth) and renewables (7.6 percent growth), the market seems to be in a tailspin, and Big Oil is already shifting its focus, with natural gas already making up 50 percent of Shell's production.[4] With the traditional energy market potentially facing a complete disruption given the emergence of renewables, electric vehicles, and battery innovations, these companies will likely pivot more toward the global plastics market for growth, even though it is a much smaller (yet faster growing) percentage of its total revenues than energy. The virgin plastic resin export market is growing in the United States due to the enormous supply of natural gas that exists in our nation. As a result, it seems unlikely that the petrochemical industry and its supply chain have much incentive for plastic recycling rates to increase from its dismal 7 percent rate that would be an impact to its top- and bottom-line growth numbers. Plastic production is becoming more efficient, which is good for the industry and us as consumers, but it will make it even more difficult for recycled material to be worked into this megascaled, efficient supply chain as a result. This is not due to the petrochemical industry being intentionally against recycling, but rather it's the nature of doing business as a dominant-design supply chain player.

The global plastics market continues to grow at a rapid rate, anticipated to reach $654 billion by 2020, essentially the same revenue of all of today's major metals combined.[5] Think about that: over a few thousands of years, we have incrementally increased the production and use of metals, only to be surpassed by this synthetic material in less than a century! Some believe the alternative to conventional plastic to be from another mega supply chain: that of the agriculture sector to provide bioplastics. According to the World Bank, agriculture is a $3.2 trillion business, almost twice as large as Big Oil. In this battle of the big supply chains, each trying to take a larger piece of this growing market, it is Big Oil trying to grow in this fast-moving plastics sector versus Big Ag, which also believes it can become a player in this space. According to *European Bioplastics*, approximately 37 percent of global land is used for agriculture, and of that land, only two tenths of a percent of is used for bioplastics, a great possibility for growth.[6] According to the agriculture enthusiasts, the descendants of Norman Borlaug's Green Revolution, we could use the growing process of agriculture to create closed-loop systems for plastic in order to offset the adverse impact of using fossil fuels or even natural gas to develop these synthetic materials. It is the fight of the two largest dominant-design supply chain systems on the planet, fighting over who will succeed the most from this growing and profitable global plastics market. The question is whether either of these dominant designs will be able to address the Peak Plastic problem that is addressed in this book.

What would David (the innovator) do after he killed the first Goliath with his slingshot if afterwards another giant Philistine was there in front of him waiting to take battle? What if the second Goliath, the bioplastic alternative that is posing as a disruptive innovation, is really from an even larger dominant design of Big Ag, was promoted as more environmentally friendly than natural gas, was less economically viable and yielded the same problems in environmental, health, and safety as conventional plastic, but was viewed as sustainable? A green feedstock to make plastic but an end product material that when transformed to a synthetic is not this green plastic it is cracked up to be. Is not this just the battle of the mega supply chains, the incremental innovations? What is needed, and what I will discuss in the balance of the book, is true discontinuous innovation, a disruption taken forward through exponential thinking that completely redefines plastic from what we may think of it today while offering all the same benefits. Not simply linear thinking in changing the feedstock from petrochemical to agricultural, but linking the industrial supply chain system to our natural one; not changing the material, but changing the entire system!

For decades of students in supply chain management, we have been taught that the growing of scale of a material and product in a supply chain always leads to efficiency, but is there not a limit to a model of growth if it becomes restrictive to our natural ecosystems, such as our oceans and our own bodies? Yet if this discontinuous model of design and supply chain is extended into nature, then the economic growth curve is in sync, and this changes the definition of efficiency and innovation, and our definition of the importance of scale should change as well, as I will discuss in the solutions of the book. Therefore, thinking of a material as an innovation is an incremental way of thinking, whether it is a continuation and improvement of fossil fuels or the replacement of them with bio-based feedback—both with the same synthetic end product result. Yet as energy markets take the course that they are on, it is only natural, and actually the responsibility of the petrochemical companies, to offset revenue gains in other markets, especially as they are growing. In the future, this will place the objectives of the environmentalists even more in dispute with those of Big Oil, which is fine as long as the innovation channels are open that will lead to change—incremental and discontinuous change. This could become a lively entrepreneurship and innovation arena that becomes the starting point for the elimination of Peak Plastic.

Closed Loop = Exponential

So far in this chapter, I have noted how dependent and addicted we are to today's largest supply chain systems, Big Oil and Big Ag, and how these markets are conditioned to make adjustments to continue to grow in response to market dynamics, such as renewable energy and other discontinuous innovations. Unintentionally, if we don't protect the innovation engines, this can lead to a *crowding out* of innovation when massive industries such as Big Oil sucks all the oxygen from the markets, making investments in new technologies prohibitive. Ironically, it may be the environmentalists who are unintentionally enabling some of this to happen through their proposed recycling programs that only mitigate the problem, holding the consumer responsible rather than the producer! Innovation can drive environmental benefits as it is also proven that good economic policy as growth is the best medicine for pulling billions out of poverty despite the collateral damages discussed in this book. Innovation also releases us from an ideological debate that pins our environmental challenges as this collateral damage when we are forced to a false narrative of having to choose one or the other. This outdated economic and supply chain model must come to an end soon, as growth without regard is not just becoming a moral issue but also an economic one as well. The ubiquity, growth, and usefulness of plastic have been driven by the present-day supply chain system, but so has its damage and unregulated toxicity.

For the past century, supply chain systems have been both the solution and the problem, even if no emphasis has been paid to the latter. As the world economy continues to recalibrate, becoming increasingly fragmented and consolidated at the same time, it is leading to a new set of challenges on behalf of its shareholders and stakeholders. Supply chain systems have been the problem and the innovation in history, and this will be the case as well in the ensuing decades, but with a twist; the discipline of supply chain management must innovate itself, stepping away from a silo model that doesn't account for the environment. Rather than a model of limits, an "end of growth," as the environmentalists state it to be, it must become a model of exponential growth, and exponential thinking, of stopping to try to save the planet or limit the impact, and to increase the impact, an impact of a *supply chain for good*.

There are no ifs, ands, or buts about it: it is the exponential thinking of innovation that must lead the discussion of how we reuse today's materials rather than a continual dependency on virgin materials with no consideration of waste and damage. The linear thinking of today's supply chain model is not capable of solving this challenge one step at a time, such as considering waste reduction as a problem with all other variables held constant. In contrast, an exponential innovation and supply chain model is one of disruptive innovation, starting from scratch with a new model rather than attempting to augment the dominant design. Rather than pinning the guilt or blame on one element of the supply chain system, such as on the manufacturer or the consumer, a disruptive innovation model must begin with material reuse, over and again, within the same closed-loop model. Agricultural and plastic waste, as well as fossil fuel by-products, should be reused over and again in this system. It becomes an anti-Malthusian curve: rather than the population growing geometrically while the production process is limited to a linear line, what if the population continued to grow geometrically while our production resources grew in a closed-loop manner to mirror an exponential representation? This becomes a new formula of "closed loop = exponential" as representing how we break free from our dependency of a supply chain system that requires us to choose between waste or no growth as a false narrative that drives us to failure.

Supply Chain Innovation for Plastic

So far in this book, I have illustrated the problems associated with the current model of plastic, and in summary it is this: over the past century we have created this wondrous innovation and headache of plastic through growing it geometrically without understanding it from an overall economic, ecological, and health perspective. Since recycling and changing the feedstock is only incremental thinking, we need an exponential mindset in order to catapult the solutions that are necessary to fix the problems in a fraction of the time. But what does it mean to think *exponentially* beyond just a concept? Is it simply enough for some author like me to define closed loop as exponential, and then let someone else worry about how this concept should be implemented? Those who snickered and smirked during my *Recycling Myth* conversations may expect as much, but as you will see in these final chapters of the book, it is altogether possible if we change our metaphor of the meaning of supply chain innovation. Starting a focus in this new approach to supply chain may be a little late, or rather a century late, but according to my forecast of Peak Plastic, we still have time to address it if we face these issues head on starting today, in 2018.

What are the breakthroughs needed to be put in place as solutions to avert Peak Plastic?

1. Solution 1: Immediately Band-Aid the Problem to Stop the Bleeding
2. Solution 2: Open-Source/Access Plastic (Open Source Capitalism)
3. Solution 3: Sustainable Polymerization
4. Solution 4: A Closed-loop Plastic System
5. Solution 5: Address Microplastics

These solutions are 21st-century answers that avoid having to choose between the economy and the environment. First, we need to stop the bleeding given what's happening today, and then we need to change the entire supply chain system from the beginning (design), to manufacturing (polymerization), and finally distribution and reuse (closed-loop system). This is a market-based approach to the plastic problem. According to the Plastic Ocean Foundation in its inquiry to the House of Commons Environmental Audit Committee in 2016, plastic should be classified as a toxin when it escapes into the environment because of its unique ability to attract poisonous chemicals "like a magnet" even if it is not poisonous on its own.[7] Such facts are important for improving the public's knowledge of the problem, but will do little, if anything, to solve for it. The best place to start in engaging the public (without relying upon it) is through creating public knowledge about these materials in an approach to develop an open-access, open-source design approach to plastic. These alternatives are not an effort in themselves to eliminate private markets but rather to enhance and save them through a paradigm shift in thinking of how these materials will be managed in their afterlife, or more importantly, in a closed-loop fashion.

One of the biggest fallacies in this field is the perception that using a renewable feedstock will solve the problem, but this is a lack of understanding of the polymerization process, including the use of additives, that makes much of the plastic what it is today. So environmental-friendly terms such as "growing plastic," "closing the loop," and "making plastic organic" are more cliché than reality until we dig in deeply to the supply chain system. Natural polymers are these materials that have been perfected over hundreds of millions of years, so should we believe that it is possible to create such novel materials in afterlife while also having such amazing characteristics in consumer use in a manner of a decade or two? I think it's possible if we reinvent how polymerization works, and how this integrates into a closed-loop supply chain system for real. There are significant changes required for this to happen to transform a conventional megascaled system to a community-based solution that will be addressed in chapter 9.

To address the challenges that I have discussed in this book, it will require more than simply a redesign of a product, but rather the overall supply chain system of plastic. The failure of a used plastic bottle or toothbrush that has escaped, hundreds of millions of times over, into our ecosphere is a much greater problem than we have ever considered, which is why the problem persists to the degree that it does in our modern world. If we continue to believe that an incremental mindset of simply collecting and reusing via recycling programs will solve for the problem, we will continue to be complicit ourselves in contributing to the problem. Simply *doing our part* is not enough; I absolutely advocate for the consumer to recycle, but not as a solution to the problem. We as consumers must understand that we are a part of the supply chain system that is leading to how and why these problems are persisting, but we cannot and must not take responsibility for the problem that we did not create. Yes, some will contend that we as consumers enable the damage, but is it reasonable to assume that anyone would reject market innovations that make our lives more habitable, such as the convenience of disposing of a water container or diaper rather than cleaning and reusing either? The transference of the responsibility to those who can do little about it has become the leading cause for why our environment is being destroyed.

Consumers are off the hook, so to speak, but it does not mean we can walk away from the problem because what's happening with Peak Plastic has and will continue to be an adverse presence in our lives whether we like it or not. Hopefully after reading this book, you will understand greatly how our lives are impacted by these global supply chains for the good, such as affordable, high-quality products made in Asia, and for the bad, such as the environmental catastrophe that makes its way to us that we often do not realize. Your shrimp may not taste as though plastic or other endocrine-disrupting chemicals has infiltrated it, and neither may any of your water sources, but it still can be the case. You may be off the hook, so to speak, but what about holding producers accountable for their products? Similar to how parents are watching how their foods are produced, toys their children play with should become a health concern as well. Like Norman Borlaug did a century ago, and German chemist Michael Braungart proposes today, we need the next Industrial Revolution to be a Green Revolution, not because we are trying to save the planet, but rather to *save ourselves*. There is no conspiracy working against this from the influences of Big Oil and Big Ag, even though they are present as mega supply chains with mega influences in their trillion dollar industries, and arriving to these conclusions is a cop-out of not understanding how markets work and the disruption that can be caused by supply chain innovations.

Michelangelo is credited with saying over 500 years ago something that has rung true ever since: "The greatest danger for most of us is not that our aim is too high and we miss it, but that it is too low and we reach it." For the last century, plastic has emerged to such a degree that our local supercenter is over 100,000 square feet in size, and has over 100,000 unique products, with a majority of them containing plastic ingredients and/or packaging. The production and distribution of a product made of plastic is so cheap because of the low cost of the material, low cost of shipping, and it was produced by low-cost labor thousands of miles from your house. These are only two of the supply chain innovations that have improved your way of life in ways that could scarcely have been imaginable just a few decades ago; could you imagine your life without them? The answer is no, and nor should you be faced with the grim options of having to choose between your comfortable way of life versus that of destroying the environment. Who would have ever believed that such a magical consumer-based economy of everything we have ever wanted, and many things we could have never imagined, would be defined as "aiming too low and hitting the target"? Today, for better or worse, we know this to be true, and we should begin to conceptualize what it would mean if the costs of Peak Plastic were eventually greater than the benefits and prevented us from using these products. But we must make a choice, and we must make it now: to not choose between consuming or not, but rather in how we design, manufacture, distribute, consume, and postconsume. These are choices that we can make that rebalance our manmade world with nature rather than a nonchoice that makes it a contentious relationship. To do so, we must view both the solution and problem of plastic much differently—are we ready to do so?

References

1. Desjardins, Jeff. (October 14, 2016). The oil market is bigger than all metal markets combined. *Visual Capitalist*. Found at: http://www.visualcapitalist.com/size-oil-market/ (accessed June 29, 2018).

2. Katakey, Rakteem. (January 25, 2017). BP sees a future of slowing oil demand growth, abundant supplies. *Bloomberg.* Found at: https://www.bloomberg.com/news/articles/2017-01-25/bp-sees-a-future-of-slowing-oil-demand-growth-abundant-supplies-iyd1s0zo (accessed June 29, 2018).
3. *The Economist.* (February 25, 2017). A world turned upside down. Found at: https://www.economist.com/news/briefing/21717365-wind-and-solar-energy-are-disrupting-century-old-model-providing-electricity-what-will (accessed June 29, 2018).
4. Paraskova, Tsvetana. (October 17, 2017). The natural gas market is set to boom. *Oilprice.com.* Found at: https://oilprice.com/Energy/Natural-Gas/The-Natural-Gas-Market-Is-Set-To-Boom.html (accessed June 29, 2018).
5. Grand View Research, Inc. (July 6, 2015). Plastics market worth $654.38 billion by 2020. *Cision PR Newswire.* Found at: http://www.prnewswire.com/news-releases/plastics-market-worth-65438-billion-by-2020-grand-view-research-inc-511720541.html (accessed June 29, 2018).
6. European Bioplastics. (April 8, 2013). European Bioplastics publishes data on land-use for bioplastics: Feedstock required for bioplastics production accounts for only a minimal fraction of global agricultural area. Press release. Found at: http://www.ccpl.it/pls/portal/docs/PAGE/NATURALBOX/STUFF/RASSEGNA_STAMPA/EuBP130408.pdf (accessed June 29, 2018).
7. Johston, Ian. (May 3, 2016). Plastic should be considered toxic once it gets into the environment, MPs told. *The Independent.* Found at: http://www.independent.co.uk/environment/plastic-microplastic-microbeads-pollution-toxic-environment-house-of-commons-environmental-audit-a7011256.html (accessed June 29, 2018).

CHAPTER SIX

Solution 1: Stop the Bleeding

Bandage on a Gunshot Wound

In an effort that was more focused on promoting socialism rather than food safety, legendary muckraker Upton Sinclair spent seven months as an undercover worker in the Chicago stockyards and slaughterhouses to gather information to write his famous novel *The Jungle*, in which he depicted the harsh conditions that immigrants faced while working in the meatpacking plants of Chicago. Being an outspoken socialist, Sinclair's primary purpose was to advocate for workers' rights and to advance socialism in the United States. Sinclair never accomplished his primary objective, but his work became the centerpiece leading to the Federal Meat Inspection Act of 1906, a landmark provision put in place as the first consumer protection act in the United States. Sinclair realized that his instant celebrity was not what he expected when he noted, "I aimed for the public's heart, and by accident, I hit it in the stomach."[1] Today, our slaughterhouses are much cleaner than they were 100 years ago, and the sheer efficiency and scale of our supply chains require as much; today's operations will slaughter 40–50 chickens in less than a minute, transforming live animals into chicken nuggets and other products out the door in less than three hours to be served in nearby restaurants by the next day. Yet in these high-volume, fast-speed operations, safety and quality incidences do occur, such as Case Farms being cited for 240 safety violations by the Occupational Health and Safety Administration (OSHA) in one year (2015), and the lack of sufficient tracing of seafood caught in the United States but processed in China prior to being sent back to the United States as "local caught."[2] For sure, we should have a greater sense of security for our 21st-century food supply chain than a century ago, and yet as the global supply chain becomes larger and more complex, greater challenges must be addressed that should raise concern to the consumer today.

Part of the reason why our food supply chain is more stable is the result of an engaged, informed public. Starting with Upton Sinclair, and continuing through today, there is a level of transparency required that food suppliers must support; today, there is no such requirement for resin manufacturers and the growing use of plastic synthetics in all facets of our lives. And yet, a case can be made, if not fully validated through research, that some of these substances could potentially be more dangerous to us than the unregulated food that we used to eat. Populists movements as early as the 1970s have sought to make this case to increase federal oversight of control, according to EPA Deputy Commissioner John Quarles in 1975.[3] After years of negotiations between the government and industry, Congress passed the Toxic Substance and Control Act (TSCA) to provide a tool for the EPA to test and even restrict chemicals deemed as leading to an unreasonable risk to human health and the environment. The goal of the TSCA was not to restrict chemicals and their use but rather for companies to disclose their new formulas and use to the government in order to address any potential health concerns to the public. This was a tightrope too difficult to walk, too delicate of a balance to be effective. As a result, the best of intentions of the TSCA became little more than a framework with no teeth; of the 84,000 chemicals in the EPA's database, 62,000 chemicals that were known at the time in 1976 were exempted from the process, with only 250 of them sought to be tested by the EPA, and only 60 voluntarily consented by the manufacturer.[4]

The public struck back. Exasperated by the lack of oversight from the federal government, some state governments have enacted their own regulations to supplement, or even displace, weaker federal standards. In response, the petrochemical industry viewed this as overly onerous to require them to adhere to numerous different, conflicting standards and is proposing a modified federal TSCA in the face of these "job killers." Meanwhile, the public is forced to watch this spectacle and decide whether greater regulations should be enacted to protect them from these chemicals or if more regulation is ineffective and just leads to fewer jobs. While this is occurring, the EPA is undergoing a slow-drip process of review, meaning that the number of chemicals that have been reviewed is not only a small percentage of the number of chemicals in use, it is being far outpaced by the number of new products introduced every year. Each year, there are an additional 2,000 chemicals added, with a review process undertaking a lengthy seven-year process at the least, and almost never yielding any restrictions.[4] At the time of the writing of this book, the EPA is undergoing a "reform process" to reduce its "regulation reach to industry" that is moving toward fewer regulations in this already limited process. Most citizens are exasperated by the back and forth ideology between the pro- and anti-regulation forces, and as a result, have stopped paying attention to these debates. Therefore, with the federal and many state governments not taking responsibility or unable to look into these matters of safety, to the extent that they exist, who is responsible for ensuring all those plastic items in your house are safe in exposure to you and your family?

Regardless of your view, we should not expect any form of government involvement to address the public need for information and concerns, like exists effectively in the food industry. Despite the best of intentions, we should also not expect there to be a change in public perception regarding the safety issues of these chemicals as a result of nongovernmental organization (NGO) campaigns; these are critical sources of information and advocacy in our communities no doubt, but there is not a broad enough constituent audience of interested citizens to listen to this message, unfortunately. My recommendation is for these groups to change their approach: instead of informing the public and advocating primarily for regulatory reform, they should focus on informing the public and becoming an engine for innovation to call out the future supply chain innovators to solve for this problem! One very effective organization that I support that takes seriously the need to inform the public is the Plastic Disclosure Project, an initiative to solicit organizations to measure their plastic waste in order to reduce, recycle, and reuse. Certainly, this is a good idea; conducting an assessment to measure and manage a company's plastic footprint can be very useful if and when the company and consumers understand why this is important. Similarly, there is another project being undertaken at the University of Massachusetts led by Mark Rossi that uses a tool to conduct an internal assessment for companies to benchmark their chemical footprint, a voluntary attempt for industry to comply to some concern that is not well understood and agreed upon. In these circles of those who are focused on this problem but may not get the attention they warrant, the question is whether our efforts can scale and improve stakeholders to the level that is necessary, and unfortunately, the answer is no. However, if these groups focused on public information disclosure and the innovation engine rather than trying to influence existing resin manufacturers and government intervention, there would be a greater movement forward to real solutions.

And yet any way you slice it, such actions are merely bandages placed on a gunshot wound: mitigating schemes that are insufficient fixes to an overwhelming predicament. There are a number of plastic resin recipes that exist that are unknown and undeterminable, possibly in the thousands or even tens of thousands, leading to a legitimate feeling of futility that anything can be done to control it. Every day in the United States, newborn babies are born with up to 287 chemicals in their blood, many of them due to their mother's exposure to various types of plastic.[5] Maybe none of these chemicals have been specifically identified as a health risk, but none of them have been empirically cleared as safe either. Are increases in heart disease, autism, ADHD, and depression causal, correlative, contributing or unrelated to our heightened increase in use of plastic and other synthetics? We don't know. What's clear, however, is that our increased use of this stuff is causal for why plastic and synthetics are found at high levels in the most remote regions of the world, on the highest mountains or the deepest depths of the Earth's crust. Today, the limitations of our public policy approach is based on an "innocent until proven guilty" philosophy, a logical starting point for the criminal justice system but a suspect policy for a synthetic material like plastic.[6] The bandage of conventional recycling, weak municipal regulations, and well-intended but limited-scope NGOs should be viewed as what they are: insufficient to even mitigating the problem of a product growing so rapidly worldwide. We need to do more, which leads us to our future of Solutions 2–5, but in the meantime we need better, more effective solutions that we can utilize in the short term, and immediately!

Better Band-Aids

Solutions 2–5 will be the game changers that will fix this plastic problem over the next 10 to 20 years through a supply chain approach that links the economy to the environment. They will address how the current system needs to be transformed, if not replaced, and must also be a shift in thinking rather than an incremental bandage. We are out of time in the most desperate areas of the world, such as the urban slums of the developing world and our oceans, but we cannot take measures that do not exist. The hope is that these short-term solutions will be suitable to put in place and need to be enacted immediately in order to stop the bleeding, or at least, to slow it down sufficiently in order to put into place more sustainable solutions.

Band-Aid #1: Build Transparency and Trust between Consumers and the Plastics Industry

Think about examples from the past, whether it is cigarettes, opioids, or other products that cause damage to consumer health and trust, and consider how it was a lack of transparency and trust across the supply chain system that led to these problems; even in the case of opioids, an already regulated product in the pharmaceutical industry, a health crisis epidemic is upon us with the root-cause problem being a lack of transparency and trust across the supply chain system. According to a poll conducted by the Mellman Group, 78 percent of Americans are concerned about the impact of toxins to their children's lives, and 33 percent believe exposure to toxic chemicals today is a serious issue.[7] On the other side of the supply chain, private companies have a right to protect their intellectual property for commerce reasons, and any form of mandated public disclosure of the bills of materials in the manufacturing process would enable greater transparency and would impact the legitimate rights of the producer, impacting trust. So what gives? How can consumers feel comfortable purchasing these synthetic products through public disclosure without damaging the intellectual property rights of the manufacturer? This dilemma reminds me of the famous quote from noted writer Stewart Brand: "Information wants to be free, but it also wants to be expensive," meaning that the cost of information in society continues to decrease and become more transparent as a result of modern technology, but information is also valuable when privately owned and must be protected. As a result, we now have probably the most important and prolific material in our society without any substantive information disclosed due to the protection of intellectual property rights.

Solving this plastic problem must commence through a contract of trust: consumers must have access to information regarding plastic products, and on their behalf, researchers should also have access to conduct more testing and experiments for the public good. It should be the obligation of the supply chain system, as should have been the case with cigarettes and opioids, to voluntarily disclose information to the consumer and researcher in a manner that ensures trust and confidence while at the same time protects the intellectual property rights of the producer. Today, this challenge is closer to being solved than ever before with a transactional/ledger technology becoming available called "block chain technology," which is a distributed, public ledger platform that can be fully secure in a peer-to-peer relationship across a supply chain. Using this technology, ledger and transactional data can be both public and private, which is exactly what is needed in order to protect the rights of the consumer as well as the producer. For example, an inexpensive plastic toy that is produced in China and distributed to the United States can have a list of *block* transactions that are associated with it where this information can be disclosed or not, depending upon the need and relationship. Let's say a conscientious mother is tech savvy, and she has an app on her phone that can use a scan feature to gain information regarding the product's ingredients at the point of sale without compromising the manufacturer's intellectual property. As a consumer, this mother may choose between two toys, one that discloses the information about the product on the toy's packaging and another that chooses to not voluntarily disclose the information. In this process, the consumer is actively participating in the process or not, and the block chain technology enables the transactional information to be disclosed to the consumer in a safe manner. Rather than a consumer having to rely on government mandates, and the producer to be burdened with them as well, the consumer drives a market decision based on what is or is not important to her. Quite similarly, a block chain transaction could provide to the consumer the percentage of a product's content of recycled material versus another that would enable a consumer to choose their level of information need, and the producer to make their voluntary decisions as well.

The beauty of this solution, enabled by technology just becoming available in the market, is the ability to ensure a trusting relationship and transaction without mandating anything, or choosing the rights of one end of the supply chain over the other. In any case like this where the consumer is empowered, market decisions are a result, meaning that if consumers are not worried about plastic products and their resins, they will not advocate for it, and producers will not have to take any action; in contrast, if consumers believe this is an important criterion regarding product safety, the market will need to respond. An electronic trace and information process across the supply chain is a 21st-century solution for where draconian and ineffective legislation has failed. Likewise, NGOs that have struggled to politic for legislative fixes can now innovate and grow awareness within their constituency groups in a much different manner. Rather than rallying around politicians, how about getting companies to agree to disclose ingredients on their packaging through the use of scan/QVC codes? Use of block chain processes and technologies will lead to a paradigm shift in the supply chain, all the way to the consumer. Supply chains will change, and a new definition of trust will be developed where the plastics supply chain will be able to both fulfill its obligation to its shareholders to take advantage of this enormous world market and make consumers and the environment happy through an electronic ledger that ensures transparency and trust.

While the block chain technology is still in its initial stages of development at the writing of this book, there is a lot of process work that can be done to rally around this cause. NGOs and other concerned citizen groups should begin the process by finding resin manufacturers and product producers who are willing to participate voluntarily in this type of program. If a sufficient constituency group is willing to back these efforts, then it is likely that innovative producers will jump on board. The technology is gaining momentum and will be ready soon, if not now, but the big question is, do enough citizens care to make the effort? I believe that through this book, I have made a case not necessarily that plastic is toxic in all its forms but rather we as consumers and researchers need to understand it better to avoid a potential next health and environmental crisis, that of Peak Plastic. The real big question, the first short-term solution or Band-Aid, is to determine whether enough consumers and citizens in the United States and around the world believe that this is an important matter to address. I'm optimistic that the answer is yes if a case can be made that is fair to both consumers and producers. As for developing nations, information transparency and disclosure is a more immediate problem and can be addressed using even less sophisticated technologies regarding the challenges of plastic and its impact on the environment, health, and safety.

The most fundamental problem we face today with plastic and its supply chain is a lack of information regarding its costs tied to its transformational benefits. Material designs and supply chains will take longer to fix, no doubt, especially across a dynamic and complex global supply chain system. Supply chains are complex and interdependent systems, and the size of the plastic supply chain is so enormous that it will not be as simple as the flow of information about it. Yet through technology that ensures trust, it seems possible that the market could determine for itself what is necessary to share while at the same time ensuring commercial growth. Just a few sparks created by the passionate few out there today can either foment a redesign revolution or, to the contrary, fizzle out as nothing of any consequence. Equally as important to the producer, it takes all the conspiratorial perspectives out of the way, as the market will drive interest or not. It is the important first step of a Band-Aid that begins to turn the tide of Peak Plastic!

Band-Aid #2: A Ban on Microbeads (and Glitter) across the World

Okay, clearly from this book you can tell that I do not place a lot of faith, based on the data, of what regulation has been able to accomplish related to plastic and the environment. As is shown in Band-Aid #1, there are better market methods to ensure information transparency and disclosure in order to let the consumer decide. Yet because plastic in the past and present has been defined by so many complexities in design and its supply chain without adequate information, it has been difficult, if not impossible, to find any traction in it from the broadest definition of its base polymers down to its micro and nano particles that are too small to even manage. To our largely untrained eyes, the plastic problem is often defined as litter on the beach, but as you've already learned from this book, the greater problem appears to be what lurks beyond our vision but is evident in our food, drink, water systems, and even present as dust that filters into our meals without us knowing about it. Therefore, an immediate Band-Aid #2 should focus on this problem of micro-plastics, not just in the oceans as it is often advertised but in our own homes as well. Micro-plastics, in general, follow three general categories of materials: primary materials, or those specifically created as microbeads, secondary materials, or those as a result of an industrial process through the use of a polymer in an abrasive activity, or tertiary that is the result of a macro-plastic leaching into small materials in the environment. The goal of this short-term Band-Aid is a focus on primary micro-plastics as the starting point.

Microbeads exist today in a variety of consumer uses, such as facial cleaners, toothpaste, cosmetics, and detergents as primary materials. Every day just in the United States, there are 8 billion microbeads that end up in waterways, a staggering number of plastic waste that does not go away or biodegrade.[8] Some states have enacted bans on consumer products with microbeads, typically rinse-off cosmetics and some toothpastes, but not others, such as stay-on cosmetics that were found to be too difficult to reformulate and various types of detergents. In December 2016, President Obama signed a bill outlawing microbeads in rinse-off cosmetics by mid-2017 but left a large gap in the law, allowing companies to use microbeads in many products such as detergents and cosmetics left on the skin. Yet concerns remain in general about the safety of personal care products such as cosmetics and toothpaste, with 84 percent of recent survey respondents favoring the government to require companies to disclose the ingredients in their products.[9] Other nations have also taken action, such as the Netherlands, the United Kingdom, and Canada, but no action has been undertaken in some of the most troubling regions of the world, particularly in the developing world. Now add glitter to the list, a seemingly harmless little bottle of micro-plastic that is typically made from aluminum and PET plastic and often of little value other than as a cheap ornamental additive to products such as shampoos, cosmetics, clothing, or even on its own for party decoration. The cosmetic company Lush has replaced plastic glitter with biodegradable materials in an effort to be more environmentally friendly, and it appears as if consumers are willing to pay for it.

For Band-Aid #2, there needs to be a worldwide ban on microbeads in consumer products across the globe, including exempted products such as stay-on cosmetics and glitter. In 2015, 5 Gyres, a leading organization focused on the problem of plastic pollution across the planet, launched the #beadfree Action Campaign as the first to raise awareness to the problem and to discourage the use of microbeads, which is just one element of the problem. How can we galvanize an educational effort to raise awareness of this problem that is invisible, but present, critical, yet unknown? Unlike controlling secondary and tertiary micro-plastics that will be significantly more difficult, and thus will fall into the longer-term solutions, this problem can be addressed immediately across the planet by consumers and producers to eliminate superfluous micro-plastics that harm our oceans and own bodies. We can work through these existing NGOs, especially 5 Gyres, to alert the public that the problem is bigger than one that pollutes our ocean, as big as this problem is, to one that is potentially polluting us as well. More needs to be done in the form of a ban of all microbeads, not just a percentage of them in certain types of products in certain nations.

Band-Aid #3: Informal Supply Chains to Reduce Waste

The final Band-Aid, and perhaps the most immediate of the three, is to address the enormous flow of plastic waste into the oceans stemming from developing nations, primarily of Asia. According to a research study completed by the Ocean Conservancy, there are 8 million tons of plastic dumped into the oceans annually, and 60 percent of it is from only five countries: China, Indonesia, Philippines, Thailand, and Vietnam.[10] According to this same study, these nations could reduce their plastic waste into the oceans by 65 percent through a $5 billion investment in conventional means of management, including conventional waste-management techniques and improved recycling techniques, investments that are unlikely to happen. Of course, the problem with these types of investments is a tragedy of the commons, the unwillingness of private entities, or in this case sovereign nations, to pony up money for a resource that isn't owned by anyone (the ocean). Perhaps our thinking needs to be turned inside out: instead of making significant investments toward cool technology that collects plastic waste after it has traveled thousands of miles, collecting plastic much slower than it is flowing in, why not create private enterprise zones on these coastal communities, particularly in Asia, for these entrepreneurial waste pickers to collect and process the synthetic waste before it escapes and, at the same time, fund poverty-ridden communities that are perpetuating the plastic into the ocean in the first place? It may seem trendy to trap this ocean plastic at each of the five gyres and to sell it to 3-D printing companies in an economically unsustainable model, or just to ignore the problem given the size of it, but neither of these efforts will slow down the increasing flow of synthetics into our oceans that is threatening our lives, literally.

Time for some math: According to the National Ocean and Atmospheric Association (NOAA) of the U.S. Department of Commerce, it would cost between $122 million and $489 million to clean up the garbage patches in the oceans.[11] The Ocean Cleanup is estimated to cost $410 million dollars over 10 years, given the current state of ocean plastic waste.[12] The problem is that both of these estimates seem to be a bit inadequate or misleading; their assumptions are based on the current state of plastic that resides in the garbage patches and doesn't account for how much plastic is already in the oceans, and how to protect the entire ocean not just the garbage patches! Seems like the only reasonable solution to the problem isn't to target the waste after it slimes its way for thousands of miles but rather to prevent it from entering the ocean in the first place. By funding entrepreneurial zones for profit where the millions of waste pickers in the world live, we stop the flow of plastic waste into the ocean, while eliminating poverty, and doing so at a fraction of the cost. Phase one of this project will take some subsidizing, perhaps by the UN ($8 billion annual budget) and/or other significant NGOs, but once Solution 3 (sustainable polymerization) is in place, these ventures will be standalone profitable operations needing no public support. In this model, it isn't the tech whiz but the waste picker who is the social and supply chain innovator! From one perspective, the waste picker in developing nations is a dehumanizing and thankless role, a nearly forced occupation that leads to a cycle of poverty. On the other hand, it is some of the most innovative and effective entrepreneurship happening in an unstructured manner on the planet! In general, this informal approach to entrepreneurship is broken into two groups, with the first as a role of waste collection provider to whisk away plastic from residential and commercial settings where no legitimate waste-management systems exist and a valorization sector to sort valuables from existing trash. In the fastest growing cities in developing nations, the informal waste-management sector plays a very important role even if its financial viability is less effective than formal, commercial operations. That is, even though they are more costly, these informal ventures are critical in serving opportunities for collection and valorization that are difficult to serve and, therefore, would not be served without them. According to a study completed by Deutsche Gesellschaft fur Internationale Zusammenarbeit (Germany) in 2011, the average cost per ton for waste management using an informal waste pickers channel is approximately $60 to $120 per ton.[13] According to the World Bank, in low income nations, the cost of waste management is as low as $2–$8 a ton for open dumping, up to $10–$30 a ton for a sanitary landfill, and these are the costs that are borne by the local municipalities. In my business model, the waste pickers would not displace existing practices of open dumping (which they are a part of already) and sanitary landfilling but rather supplement it, at a much lower cost than is noted above ($60–$120). In fact, the community-based closed-loop system that I discuss in Solution 4 is robust enough, even at the start, to be on par with the existing cost per ton in low income nations. This means that if these low income nations are unable to supplement the additional tons of plastic waste that escapes their systems today, it could be funded by other means, such as the setup of enterprise zones or a NGO or UN subsidization; given the plight of our oceans, it would be an easy business case to validate!

According to the *Environmental Science and Technology* journal, a staggering 1.5 million of the annualized 8 million tons of plastic waste (20 percent) enter the ocean through one river, the Yangtze River in China; this would be the obvious starting point for this program.[14] If presently in this region, it costs $40 a ton for waste management costs that are avoided by illegal dumpers that the Chinese government is seeking to control, and a waste picker typically receives $14 a day for his job. Instead, we could implement a community-based supply chain system for these waste pickers to earn a sufficient living.[15] This innovative approach to a community-based supply chain (with the longer term solution as Solution 4) would save millions of dollars and plastic waste from ever entering the ocean, where it is doubly expensive to fish it out. Furthermore, it provides economic relief to the poorest of the world and can be replicated sequentially to the rivers of the world that are responsible for 90 percent of the plastic waste entering the oceans.

The hope is that if these three stopgap Band-Aids are put into place right away, it will slow down the trauma of Peak Plastic sufficiently for us to be able to solve this problem by 2030. These would not replace such conventional approaches as existing recycling programs or legislation, but it would be advisable to invest funding in these rather than the new investment in the former because these build upon the longer term solutions of Solutions 2–5. These bandages will address consumer awareness of the problem via transparency, an elimination of micro-plastics (the greatest issue with plastic), and the damage in the developing world where most of the problem resides. By stabilizing the situation, we set ourselves up for solving the problem with Solutions 2–5; but first, we need stop the bleeding that's happening today. It took a century for this complex problem to grow, and it grew exponentially, both in its costs and benefits; therefore, we need to first clear the field to prepare for the innovation, and then make it happen.

References

1. Bloom, Harold. (2002). Upton Sinclair's *The Jungle*. New York: Infobase Publishing, 11.
2. Grabell, Michael. (2017). Exploitation and abuse at the chicken plant. *New York Times*. Found at: https://www.newyorker.com/magazine/2017/05/08/exploitation-and-abuse-at-the-chicken-plant (accessed July 25, 2018).
3. U.S. Environmental Protection Agency. (July 10, 1975). Quarles testifies on the need for toxic substances act. *Press Release*. Found at: https://archive.epa.gov/epa/aboutepa/quarles-testifies-need-toxic-substances-act.html (accessed June 29, 2018).
4. Scialla, Mark. (June 22, 2016). It could take centuries for EPA to test all the unregulated chemicals under a new landmark bill. *PBS News Hour*. Found at: https://www.pbs.org/newshour/science/it-could-take-centuries-for-epa-to-test-all-the-unregulated-chemicals-under-a-new-landmark-bill (accessed June 29, 2018).
5. Houlihan, Jane, Timothy Kropp, Richard Wiles, Sean Gray, and Chris Campbell. (July 14, 2005). Body burden: The pollution in newborns. Environmental Working Group. Found at: https://www.ewg.org/research/body-burden-pollution-newborns#.WvzbKUgvzIU (accessed June 29, 2018).
6. Kollipara, Puneet. (March 19, 2015). The bizarre way the U.S. regulates chemicals—letting them on the market first, then maybe studying them. *Washington Post*. Found at: https://www.washingtonpost.com/news/energy-environment/wp/2015/03/19/our-broken-congresss-latest-effort-to-fix-our-broken-toxic-chemicals-law/?utm_term=.4ef3494dc779 (accessed June 29, 2018).
7. Nolan, Jamie. (September 14, 2010). New polling data indicates overwhelming public support for chemicals regulation. *Saferchemicals.org*. Found at: https://

saferchemicals.org/newsroom/new-polling-data-indicates-overwhelming-public-support-for-chemicals-regulation-2/ (accessed June 29, 2018).

8. Kaufman, Alexander. (May 23, 2016). Obama's ban on plastic microbeads failed in one huge way. *Huffington Post.* Found at: https://www.huffingtonpost.com/entry/obama-microbead-ban-fail_us_57432a7fe4b0613b512ad76b (accessed June 29, 2018).
9. Mellman, Mark. (May 17, 2016). Time to regulate cosmetics. The Mellman Group. Found at: http://mellmangroup.com/time-to-regulate-cosmetics/ (accessed June 29, 2018).
10. Schiller, Ben. (October 12, 2015). Most of the plastic in the ocean comes from just a few polluting countries. *Fast Company.* Found at: https://www.fastcompany.com/3051847/most-of-the-plastic-in-the-ocean-comes-from-just-a-few-polluting-countries (accessed June 29, 2018).
11. NOAA Office of Response and Restoration. How much would it cost to clean up the Pacific garbage patches? Found at: https://response.restoration.noaa.gov/about/media/how-much-would-it-cost-clean-pacific-garbage-patches.html (accessed June 29, 2018).
12. Slat, Boyan. (2014). How the oceans can clean themselves: A feasibility study. The Ocean Cleanup. Found at: http://www.theoceancleanup.com/fileadmin/media-archive/theoceancleanup/press/downloads/TOC_Feasibility_study_lowres_V2_0.pdf (accessed June 29, 2018).
13. Gunsilius, Ellen, Bharati Chaturvedi, and Anne Scheinberg. (2011). The economics of the informal sector in solid waste management. CWG Publication Series No 5. Found at: https://www.giz.de/en/downloads/giz2011-cwg-booklet-economicaspects.pdf (accessed June 29, 2018).
14. Schmidt, Christian, Tobias Krauth, Phillipp Klöckner, Melina-Sophie Römer, Britta Stier, Thorsten Reemtsma, and Stephan Wagner. (April 2017). Estimation of global plastic loads delivered by rivers into the sea. *EGU General Assembly Conference Abstracts* 19: 12171. Helmholtz Centre for Environmental Research. Found at: http://www.ufz.de/export/data/2/149396_EGU_2017_Schmidt.pdf (accessed June 29, 2018).
15. You, Li. (February 15, 2017). How China's garbage goes from cities to rivers. *Sixth Tone.* Found at: www.sixthtone.com/news/1942/how-china's-garbage-goes-cities-rivers (accessed June 29, 2018).

CHAPTER SEVEN

Solution 2: Open-Source/ Access Plastic (Open-Source Capitalism)

Commons Sense?

The 2012 book *Standing on the Sun*, written by Christopher Meyer and Julia Kirby, asks, "Can capitalism be adaptive in the future?" The book explores whether start-ups around the world will be bounded by old rules of capitalism that are uniquely different from experimental models, such as an open-source approach to design and capitalism. Could a new model of earning a profit and sharing exist in a capitalist model? This is an important question from the start as we determine how plastic can become sustainable in the future due to its failure to be so in a 20th-century model of capitalism. In the traditional model of capitalism, intellectual property rights are very discrete and sacrosanct, with an inventor not required to disclose any information about this product to the public unless it is proven, without a shadow of a doubt, that this product is harmful to the public; this is the challenge that I have addressed in this book with the plastic problem. While I do not expect this proprietary model of intellectual property to be changed, I do expect capitalism to flourish in new ways outside of these definitions, such as a hybrid model of public-private ownership that could address property that is owned by nobody or everyone, such as the commons. Therefore, while solution 2 is an open-source, open-access approach to design and capitalism that is not the ending of proprietary property ownership in capitalism and markets, it is an emergence of a new model that could be a more practical approach to plastic in balancing the economy, health, and the environment in the goal of averting Peak Plastic.

A well-known depiction of what's wrong with today's plastic supply chain system is embodied by a concept first presented in 1968 by American ecologist Garrett Hardin termed "the tragedy of the commons." This problem is most often referenced to how private interests can avert responsibility when public damage occurs as to their doing that can be extended to the challenge of how to take responsibility for elements of our planet that are governed by nobody specifically, such as the air and the oceans. Hardin's essay not only solved the "commons problem," it popularized it in a useful manner when thinking about environmental problems.[1] Yet when our modern version of capitalism originated in the late 19th century with the Industrial Revolution, it was still largely assumed in an economist's omission that the planet was sufficiently large and robust enough to handle any manner of industrial actions, so a problem of the common was never considered. As a larger percentage of a larger world population falls into a prosperous condition, it becomes a math problem of 7 billion on a finite planet of limits, pushing our commons areas, like oceans, and the overall planet to the brink. According to Dr. Hardin, there are two solutions to the problem: to regulate the commons or to convert it to private property. In the 50 years since Hardin's essay, many have tried to regulate the commons as a solution to the problem, only for these regulations to not be successful, or solely mitigations, and it doesn't seem feasible to privatize the ocean. Alternatives, such as the interjection of international agencies to address the local problems of traditional communities have become an even greater failure. Elinor Ostrom from Indiana University won a Nobel Prize in Economics by identifying the damage created when international donors and NGOs supplant local institutions in seeking to solve such problems as forests and rivers.[2] According to hers and other research on the matter, a "tragedy of the commons" problem may not be as much of the failure of private market, per se, but rather a poor implementation of private markets and strategies. Perhaps said differently, the commons does not need to be disposed of due to its inability to balance economic and environmental matters fairly, but it must reform itself to the realities of the 21st century that brings greater challenges than did in the 19th and 20th centuries.

This brings us to Solution 2, an open-access, open-source approach to plastic design, or perhaps said differently, open-source capitalism. A tragedy of the commons is not necessarily a failure of private markets to public interest but rather an incomplete information model across the supply chain system. For plastic to be a material that is used in a closed-loop system, there needs to be more complete information across the entire supply chain, which doesn't exist today. If a private enterprise model enables a transactional process that restricts data across the supply chain (e.g., the ingredient recipe of a plastic product), it will be difficult, if not impossible, to determine how to close the loop with these materials, technically, culturally, and economically. However, if these transactions can protect both the public and private interests collectively in the design of a material and supply chain processes through collaboration, technologies such as block chain, and a capitalist model, then everyone wins. Market activity is for profit, but it must be ensured to disclose information to the public to ensure that this transaction will do no harm. Another outdated tenet of capitalism is the assumption that markets are self-regulating due to resource scarcity, and therefore, the principles of supply and demand are in check to ensure efficiency. Today, the plastics market is near infinite, as it continues to grow faster than any other commodity; therefore, a new model for market activity must be developed that doesn't assume scarcity and no impact on the environment to one that instead assumes a need for balance within the unlimited of industry to the fragility of nature. In this model, better designs are achieved through the disclosing of information, not restricting it. Private and proprietary designs have been ignorant of its damage to the commons, therefore providing a poor model that capitalism cannot sustain. In contrast, a self-regulating model between capitalism and the environment is an open-sourced, open-accessed design model. Through new technology and open-sourced, open-accessed information, it will win in a market model versus the private-ventured, proprietary plastic design of the past century.

In the future, private/proprietary transactions and designs can exist as a function of free-market capitalism, but if these economic models are based on resources that are scarce, then it will fail in addressing the "tragedy of the commons" and the impact of plastic to the environment. Therefore, closed, proprietary models in this manner must either become more open or bear the costs of its transactions to the public property. As an alternative, open-source, open-access, transparent design and supply chains where private enterprise is protected through block chains enable both consumers and producers to make optimal decisions while not restricting market activity but actually enabling it. Take for instance, the design of plastic products, including water bottles, with the challenges of BPA; according to some chemists, it is easier to state the concerns with BPA than it is to find a safer, more functional alternative to it. In an open-source model, innovators would be financially incented to work collaboratively and transparently to develop a plastic manufacturing method that is safe through a BPA alternative. Transparency and trust can be ensured in a market model, while today a BPA alternative is simply a marketed replacement without understanding if it is a safer alternative, leaving consumers to wonder whether the product should be considered safe or not.

An open-source model of plastic design and its supply chain should not be considered as a replacement to the current state but rather an alternative to it. In a free market model, consumers should have the opportunity to make rational (or irrational) choices, and producers will need to respond to these signals. New technologies and design methods will not displace conventional producers and supply chains, especially if demand signals between them continue to remain strong. However, the metaphor of how a supply chain services its consumers will change, and these changes will provide a new model of consumerism for us to consider. I am actually very confident that when provided the right information, the consumer will make more rational decisions, and the perception of how shallow consumers are relative to making the right choice will be viewed differently. As a direct result, the megascaled plastic supply chain system will need to change accordingly: to adapt or go extinct. To paraphrase a question from the authors of the book *Standing on the Sun*, can capitalism be adapted to ensure its own survival? Rather than thinking of open-access and open-source as a threat to capitalism, in some industries, such as plastic, it might be considered as a savior for it.

Design Revolution

You might be thinking, "What a great concept, but when does it start?" My prediction is that in the next decade, the decade of 2020, we will face the greatest supply chain revolution of design-thinking of *things* in modern history. Supply chains, to some extent, have been around for centuries, if not millennia. As early as 200 BCE, the Silk Road was perhaps our first global supply chain, a 4,000-mile route where Chinese silk and tea were traded for spices, nuts, jewels, and other items across the Middle East and Europe. What was fascinating about this global supply chain system is that even though it was in balance to some extent, and capitalistic, there was a lack of transparency and knowledge from one end to the other. For instance, those in Europe had no knowledge of what people were like in China, and vice versa, and the system was often in conflict in the middle sections (e.g., the Middle East) that led to disruptions unknown. Therefore, the design of products and the supply chain were of crowds and collaboration but without any trust or transparency. Every step in the supply chain was a private transaction from one party to the next, and to the next; there was no optimization of the entire supply chain system built through trust and transparency. Even today, over two thousand years later, we are limited on building *really good supply chains* due to these limits of trust and transparency. We may know more about the products that are made today in China and shipped to the West than we did when China shipped its products across the Silk Road, but we remain largely ignorant of much about the plastic toy through the marvels of today's global supply chain system. So I quote the phrase: "the more things change, the more they stay the same."

Why not an open-source, open-access model to plastic? Last summer (the summer of 2017), my research team and I entered a competition called Imagine Chemistry sponsored by AkzoNobel, a $12 billion specialty chemical company, in an effort to open up synthetic chemistry designs to the innovation community. This progressive, multinational Dutch company has commenced the competition to bring about transformative change within its organization to solve both market and community problems, and it attracts market innovators to do so in this space. Unlike a conventional model that assumes a company of this market position will only grow through a closed, proprietary model, AkzoNobel reached out to the world of innovators in seeking to solve market and environmental challenges that they see in their business model. Other serial entrepreneurs and large companies have gotten behind the concept, such as the XPRIZE Foundation, where bold and audacious goals are launched to solve a market failure that would otherwise be considered beyond the private enterprise community to solve within a reasonable time frame. Using a common, open-source platform, with funding from a foundation, NGO, corporation, crowdsourcing, or all the above, innovators will self-organize to form a community to solve problems in a manner much different from designs of the past—traditional polymer science. Either through a prize competition or an open-source research foundation of some sort, a collaboration of innovators will complete the mapping and then design a plastic resin/product that can be used as an open-source product design rather than a proprietary formula. Rather than being solely optimized for private profit, the resin/product design will be open to for-profit companies that will be required to use the design in a closed-loop plastic system with very clear metrics of what this means, and penalized if they do not deliver these targets. In this model, a bottled water or toy manufacturer may be licensed to use a plastic recipe for a cost or even for free as long as the product meets certain environmental, health, and safety standards for a supply chain in a closed-loop system, which will be addressed in Solution 4. In this market-based model, a social innovator would be incented by both profit and conviction to develop plastic resins that are nontoxic and reusable in a closed-loop system, manufacturers will be incented by their own markets—and potentially by the freeware nature of the resin—and consumers will finally be provided a sustainable choice they can feel good about. As an example, the consumer demand for electric cars has risen remarkably in the last few years, making the case that with a viable business model, consumers will make the rational and even ethical choice.

Eight years ago, I finished my doctoral dissertation on this idea called "generative customization," a concept where designs were digitized down to the material element and all the way through the supply chain system. That was back in 2010; since then, the Materials Genome Initiative has taken root to use software, supercomputers, and screening techniques to map materials in order to improve capabilities from design to reuse. Yet the transformation has yet to happen: will it be the block chain technology that will take it forward to allow decentralized partners to work profitably and efficiently with one another to transform our existing definitions of design? Back when I was a student, I could not have imagined how these technologies would have the potential to enable the supply chain system to address these challenges that we face today. Inevitably, this will catalyze us toward a new way of thinking about designing and using materials and products, but first we must make the design process open and collaborative in order to *design for good*.

Design Growth for Good

It is a false narrative that environmental, health, and safety issues, such as plastic waste in the ocean, and potential human health problems, such as endocrine disruption, must be solved by consumers paying a premium, choosing inferior products, or the banning of products altogether from use. This misconception has been fostered now for decades that innovation and efficiency are mutually exclusive from our supply chains to our natural systems and we must grow through waste. A *design-for-good* model can be discounted when it is believed that today's closed, proprietary design and supply chain systems cannot achieve these objectives at a price and quality level that today's consumers demand. Yet our *post-scarcity* economy of unlimited material will not only make this possible but make it required. In his 2018 book *The New Human Rights Movement: Reinventing the Economy to End Oppression*, Peter Joseph notes the five attributes of automation, open-access, open-source, localization, and networked digital feedback as the foundation for a new economic model.[3] All these attributes are based on the same supply chain model that I am advocating in this book, including an open-source/access system (Solution 2), and a community-based supply chain system (Solution 4). Yet these concepts are neglected today in our efficiency- and scale-based supply chain systems that leave no other options to consider. Yet as our supply chains are failing to protect basic human interests, including those of the environment, a design revolution becomes possible and required. Ending an obsession of the existing-state supply chain model that squeezes every ounce of efficiency and scale from a process through a narrow definition of these terms is starting to leave no oxygen available for the innovation process, and this will change in the future.

What will a design revolution look like in this future? As is the case in any innovation model, it is unlikely to be harnessed by the dominant-design players in the market such as BASF, Dow/DuPont, Sinopec, and other major-producer in the plastic supply chain system. These companies simply cannot afford to take their eyes too far off of their existing customer base, and they lack an incentive to disclose their product formulations, create open-source/access designs (in their opinions), and close the loop of today's material flows with how their supply chains have been designed for scale and efficiency. Mature industries lack an incentive not because they are greedy and soulless but rather because this is what their stakeholders require them to do rather than invest in new product development that will not offer a short-term value gain. As start-up, open-source, open-access innovators enter the market, much like how Big Auto responded to Tesla, they will respond in their own manner, perhaps developing their own incubators to build these new plastic design and supply chain models. Today, especially with petroleum demand falling, the large petrochemical companies have little incentive to reintroduce recycled content into their supply chain systems, especially given how these recycled plastic markets are crashing due to a lack of stable economics. Start-ups and open-source design platforms will face scale challenges of how to gain market share versus these challenging market conditions, but they possess great opportunities to overcome these challenges through the consumer buying into their stories of less toxic waste to the environment and our bodies. The future of plastic design will neither be anonymous nor incremental; through an open-sourced approach to design, the recipes must be understood by producers of the resin, manufacturers of the products, retailers, distributors, and consumers, as well as those who will determine its post-use and researchers who will determine its impact and potential in the future.

The social innovators will have the enormous benefit of not just computer-based design as a framework but a new revolution in computing to include the Materials Genome Initiative, supercomputing (via quantum computing), artificial intelligence, and block chain technology to even the playing field to the dominant, proprietary design. In this new model is a materials genome approach that will design materials by digitizing its attributes, like exists in our own human genes. For example, a housing manufacturer may be seeking a new design to use less-controversial materials for its water pipes than PVC, a material known as the "poison plastic." This manufacturer may provide the functional specifications for the water pipe to a group of social innovators to develop novel designs. Rather than a limited universe of prospective materials and designs as exists today as being only slight derivations of existing plastics, a larger variety of options can be identified, including hybrid or even novel materials including no plastic at all. Today, the Materials Genome Initiative has already identified 69,640 inorganic compounds and 21,954 molecules, and it continues to identify new materials every year.[4] In its process, designers use supercomputers to identify the proper material to meet the specification as well as to virtually test the properties of the materials before they are ever synthetized in a physical laboratory setting. As I noted in the generative customization approach to supply chain, this will significantly reduce cost, time, and risk in the process, achieving a cradle-to-cradle design approach within the requirements needed to meet the Peak Plastic problem.

Using this new paradigm to design, not just in the material itself but the overall supply chain system, ensures that this material of plastic that has been so useful to us for the last century will continue to be so, while at the same time not reaching a Peak Plastic tipping point where its usefulness is outweighed by its detriments. While plastic as a material has been in a closed design system ever since its onset, it is time to consider its benefit and detriment within a bigger supply chain, that of a conjoined synthetic and natural system of how we must view progress in the 21st century and beyond! As such, we are able to fulfill Dr. Michael Braungart's vision of *design for good* on behalf of the economy and environment.

21st-Century Materials: Blurring the Line between Natural and Synthetic

If these social innovators and supply chain systems get it right, and we are all depending upon them to do so, they will achieve paradigm shifts in thought and execution that we will scarcely believe is possible but can happen sometime soon, hopefully in the next five years. The most important design paradigm that must be achieved, and will lead to both a materials and supply chain revolution, is a blurring of the lines of distinction between what is natural and artificial, or synthetic. For example, through advances in fields such as biotechnology, biomedical engineering, molecular genetics, and artificial intelligence, to name the most influential, is a blurring between nature and technology, and the future of our plastic must follow the same path. In medicine, it is happening with artificial body parts and organs already inside us, and in the future, some theorists believe it to be fully plausible that we will have artificial intelligence embedded into our bodies, changing the nature and definition our consciousness. In the future, it must become less discrete between an organic and synthetic substance as a result of the latter being able to be present in nature and leading to its improvement rather than its detriment. This will be the most important design element for plastic to save itself from this Peak Plastic crisis in the future.

A real design revolution will be completely fulfilled when our natural and synthetic substances are synergetic with each other and our bodies across both our natural (ecosystem) and industrial supply chain systems. Hopefully one day, humans will laugh at these recycling routines as the collection, cleaning, processing, and converting of industrial materials separated intentionally from nature. No longer will plastic be this foreign material of both good and bad, but only good. This mysterious material, originated from the remnants of fossilized plants and animals, processed over tens of millions of years, will be designed to fit within a circular system, of not just nature but our 21st-century industrial model as well. For this to happen, a sound baseline of knowledge must be available to the public regarding today's plastic supply chain in order to transform this material to a design for good, something I believe that we all would like to have happen as soon as possible. Novel materials in an open-source model means that solutions will no longer be restricted to a certain feedstock, such as a petrochemical feedstock with dangerous additives, but rather can use seemingly unique and unknown materials, such as plastic was in the early 20th century. If the platform is built collaboratively with knowledge and an open-source approach, anything will be possible as a function of supply chain innovation.

It's important to understand the role of the transaction in the supply chain system; starting with the Silk Road and leading to the promise of a peer-to-peer, almost infinite, and completely secure transactional ledger system, the market will gravitate to an open-source, open-access approach to design profitably in a closed-loop system. Collaboration among universities, NGOs, and other impartial researchers who are interested in forwarding knowledge on the topic is the building block to materials that can live well in our natural and synthetic worlds. Developing an open-source portal with these goals should lead to an acceleration of innovations across new resin and product designs, as well as research findings on critical topics such as plastic's impact on our oceans and our bodies, as well as novel ways to addressing challenges. There could also be an opportunity for product developers to reverse engineer existing plastic recipes into new formulas that address problems. Through this endeavor that aspires toward knowledge, there are possibilities of collaboration, innovation, and transformation that could lead to discoveries that are not possible in today's fight between mandates and regulations versus proprietary trade secrets and massive, complex, monolithic supply chain systems. The possibilities are literally limitless through the power of an open-sourced, crowdsourced, and social media–driven approach to addressing the one real problem that we should all agree to regarding plastic: today, we do not know enough about it.

Deformulate Existing Design

We should not forget that while information wants to be free, including in a for-profit, private-enterprise world, it is also necessarily expensive, and the role of intellectual and private property won't and shouldn't change this much in the future. Yet the friction that exists in a closed, proprietary system will begin to be reversed as social innovators uncover these designs and place them out in the cloud for responsible private-enterprise use. The social innovators could also begin to act as hackers of some sort, not in an effort to steal intellectual property but rather to enhance existing design, an inevitable dissonance between the public goal and the private property. Deformulation is a chemistry-based reverse-engineering process to determine the components of a specific product. Much like the goals for intellectual property that are in place to advance the collective knowledge, deformulation seeks to do so as well through an education and competition process. Not to get too technical, but there are a variety of techniques that can be chosen to conduct these experiments, such as gas chromatography with mass spectroscopy (GC-MS) or 1-D and 2-D NMR. Any type of sequential deconstruction or depolymerization will work through the most labile to stabile bonds and will be particularly strong in identifying additives without providing the footprint for the steps to produce the polymer. Therefore, an open-access, open-source model within a social entrepreneurship community could effectively replace the need for the EPA to study and test a few materials over the course of many years in a model with too limited of capacity of just a few chemicals out of the thousands, and little hope of an education process for plastic. Based on a pure market model rather than costly, political, and inefficient government intervention that is parsed from nation to nation, consumer insights and involvement can be studied and funded in a much better model in the future. To ensure the integrity of the process, multiple experiments can be conducted by different chemists in a crowdsourcing manner to ensure that there is consensus in the design of the resin or product.

Once completed and validated, the findings of the deformulation process can be provided to all consumers and private-industry producers transparently and respectfully through effective open-source knowledge portals. Researchers can use the data to conduct further studies in an effort to further expand our knowledge of the plastics market. Other social innovators can use this data to develop new formulas in the hopes of improving them, and possibly even making their findings open-source as well. This could lead to an innovative approach to plastic supply chains where new designs are formed that transform the entire supply chain system, as will be discussed in the next solution. It could also lead to a push for dominant-design players to voluntarily disclose their recipes as an act of good marketing rather than waiting for their formulas to be deformulated. The initial goal for this solution is to advance knowledge regarding existing designs, improve public knowledge, and enable improved research, but it will ultimately lead to product and supply chain redesigns that optimize this new paradigm of *design for good* that improves the economy, environment, and our health.

Knowledge, Now Label

Social innovators take note: we have undertaken the most critical step in the long-term and profitable viability of plastic as a material in a supply chain system. Putting the ball in your hands will enable you to lead change in an industry that was not seen as possible, to drive change much further than through recycling programs or government regulation. As these new resins and products are developed, the private enterprise becomes accountable for not only ensuring the product follows a fully closed-loop supply chain system but also that product labeling for complete transparency is provided to consumers. This will not occur through a mandate such as the FDA Nutrition Labeling and Education Act of 1990 (NLEA), but it will be equally as effective because it will also be driven by consumer advocacy and this time in a 21st-century manner! Prior to the 1960s, the only food labels that were provided were voluntary labels pertaining largely to diet restrictions, and there was less of a need for them because most meals were prepared at home and made from basic ingredients. However, with an increased understanding of the relationship between health and diet, consumers increasingly wanted to understand what was included in their food, particularly as more became processed products. After a few attempts at regulation, the FDA implemented NLEA, which proposed rules for mandatory labeling for almost all packaged foods. One study found that the implementation of the NLEA law by itself did not improve the diet quality of Americans, the use of labels did.[5] This solution will accomplish the same objectives as the FDA NLEA but through an open-source, open-access system. We will have accomplished what is necessary for consumer education through the use of innovation, markets, and supply chain. For resins and products not voluntarily disclosed by producers, there is an opportunity to ethically hack or deformulate them, and then provide this information available to the consumer in the public ledger block chain as well. In this market condition, producers will not be required by law to offer transparency in design and fulfillment but rather will do so as a matter of demand from the end consumer. When you wish to purchase a reusable water bottle at your local Target or Walmart, you will have disclosed information, or not, and if the information is not disclosed, you can snap a picture of its bar code to either research the product through your smartphone, appeal for social innovators to hack it, or to not purchase the container altogether given a lack of transparency. That way, you won't be tricked into thinking because it is "BPA-free" that it is assumed safe simply because you know nothing more about it. Think about how this will redefine our knowledge of plastic as consumers, and our purchasing decisions as a result. An expecting or new mother will make better, more informed choices on behalf of the health of her baby rather than essentially being in the dark as she is today. Much like with food, it should be the consumer's right to make an informed decision in a modern society.

Imagine the possibilities after this solution is enacted, and how it can lead to a paradigm shift of how plastic is design, manufactured, used, and reused, as well as its impact on us, society, and the environment. And with this now in place, we can move forward to even more exciting solutions to rid the world of a problem like Peak Plastic.

References

1. Adler, Jonathan. (May 22, 2012). Property rights and the tragedy of the commons. *The Atlantic*. Found at: https://www.theatlantic.com/business/archive/2012/05/property-rights-and-the-tragedy-of-the-commons/257549/ (accessed June 29, 2018).
2. Tierney, John. (October 15, 2009). The non-tragedy of the commons. *New York Times*. Found at: https://tierneylab.blogs.nytimes.com/2009/10/15/the-non-tragedy-of-the-commons/ (accessed June 29, 2018).
3. Joseph, Peter. (2018). The new human rights movement: Reinventing the economy to end oppression. Dallas, TX: BenBella Books.
4. Materials Project. (2018). About the Materials Project. Found at: https://materialsproject.org/about (accessed June 29, 2018).
5. Asirvatham, Jebaraj, Paul McNamara, and Kathy Baylis. (July 25, 2010). Did implementing Nutrition Labeling and Education Act (NLEA) of 1990 improve diet? Selected paper prepared for presentation at the Agricultural & Applied Economics Association's 2010 AAEA, CAES & WAEA Joint Annual Meeting. Found at: http://ageconsearch.umn.edu/bitstream/61660/2/11974_Asirvatham_McNamara_Baylis_NLEA1990paper.pdf (accessed June 29, 2018).

CHAPTER EIGHT

Solution 3: Sustainable Polymerization

The Polymer Problem

Polymers are the story of life. They are the DNA and proteins in our bodies; the cellulose in plants that is produced in a water-based, ambient-temperature environment; and fully degradable and reusable. How polymerization started on Earth is not clear, a miracle performed and perfected over hundreds of millions of years. For sure, synthetic polymerization has been one of our greatest innovations, but one with unintended circumstances that requires massive amounts of nonrenewable fossil fuels and chemical additives as an input, and unnatural, leeching remnants as an outcome. The polymerization process is the link between the organic and synthetic polymer, the same process, at least in general terms, as has been responsible for creating, reusing, and re-creating the same organic materials in a cycle of nature for billions of years. Therefore, to solve the problem of plastics, the inputs are less important than the process of how the outputs are created, to develop them more as they exist in nature.

Man's first use of polymers occurred through harvesting nature, plants, and animals such as spider silk, rubber tree extract, tortoise shells, and so on. These materials were useful but limited in design and scale as first-generation materials for human civilization. If our use of these materials grew with the population in the 19th and 20th centuries, we would have certainly destroyed nature through overharvesting to keep up with population levels and still been deficient in our material requirements. In the 21st century when our world population may tip to 10 billion inhabitants, our polymerization strategy will be critical to not destroying the planet and ourselves through this use of unnatural synthetics. Yet the greatest problem may be in not realizing what the real problem is, the problem of manmade polymerization, the creation of plastics at high pressure and temperature that is resistant to nature while leeching toxic chemicals, the very definition of a Frankensteinian process, yet performed at such scale and complexity to be even more terrifying in actuality. Solving for this problem won't be simple in our mass-scaled petrochemical supply chain process that is embodied through use of high pressure and temperature to make an almost unlimited array of plastic designs. The synthetic chemistry industry is big business; the worldwide chemicals industry is massive, closing in on annual revenues of $4 trillion, and growing to $6.6 trillion by 2035—a sizable market opportunity—according to research from Roland Berger. Today, the natural polymer market is less than 1 percent, not even a point of discussion in today's market.[1] Can natural polymers take a larger piece of this growing market, given the polymerization process, or is something else the transformative solution? Technology could make it possible, as I will discuss in this chapter, but it will need to disrupt this large dominant industry that is already being shaken by the electric vehicle and renewable energy industries. The question is this: How will the petrochemical supply chain system respond to disruptions already happening in its sector if a further disruption occurs through changes in the polymerization process, and a new definition of plastic?

The Recycling Distraction

And then there's the distraction of recycling. I am a steadfast believer that we all should recycle, but only as a mitigation, not as a solution. We cannot recycle our way from out of this plastics problem due to this unnatural polymerization process that defines synthetic plastic within its supply chain system. The best dreams of recycling programs across the world would at best be able to keep pace with the growth of this product, therefore keeping constant the tens, if not hundreds, of millions of tons that are dumped into the environment every year. Whenever this topic comes up in debate, I refer to the polymerization process as the problem, and not the input material that is often viewed as the case. In any legitimate use of the term "closed-loop system," which will be discussed in more detail in the next chapter, the act of collection must be followed by de-polymerization and then re-polymerization in order for it to occur, by definition in a like-for-like plastic reuse system. In today's recycling programs, the costs of collecting, cleaning, and processing is so high in comparison to the value of the used material, the only conversion process able to be implemented is something called "mechanical recycling," where the used material is cleaned and ground in order to be reformed again. In fast-growing regions of the world, such as Asia, the supply chain and conversion costs are too high, and through using this inadequate process of mechanical recycling, the usable material yields are too low. Because using a superior conversion process of de-polymerization, or restoring the monomers back to what they were originally as an input, is not economically feasible to the overall supply chain costs and the value of the used material, it cannot be utilized. Instead of this as a failure point, how about turning this supply chain inside out and focusing on depolymerizing the polymers, and then working to re-design the supply chain system in order for it to be viable? The epicenter of the problem is how these materials are polymerized, a problem that is almost never mentioned when the advocates educate the public on the importance of recycling. After this is addressed, we can move forward with a new approach to supply chain.

Not many people really understand that it is a *polymerization* not a *recycling* problem with plastic, its design and use; recycling as a strategy, not an action, becomes a distraction to the real problem. The most common approach for recycling/reuse in plastic is a method known as mechanical recycling: the process of grinding, washing, separating, drying, re-granulating, and compounding the material back with virgin material.[2] In this process, the used plastic is grinded down in order to be reformed and reused; the material is not depolymerized, making it a more economical approach but also less efficient (lower yields) and certainly less environmental. In the PET market, the recycling rate using mechanical means is approximately 15 percent of the total used material, despite efforts made to increase recycling rates; this means that less than a fifth of used PET is reused back into its original material stream. In nations in Northern Europe where special care is taken to separate materials in a very costly process, and high percentages of material is collected and recycled, these used containers can achieve a material recovery yield of up to 50 percent or even higher, therefore reducing the amount of new PET material that must be received from fossil fuel–based material. In other nations with less pristine methods of recycling, the yield rates are much lower, where recyclables are mixed with trash and contaminated. One of the most successful mechanical recycling plants in the world is in South Africa, in the city of Johannesburg. Here is Africa's first bottle-to-bottle recycling plant run by Extrupet, a large-scale operation that produces 14,000 tons of recycled PET, or rPET, in a year. This operation required a $6 million capital investment just for the conversion equipment, and probably double that amount for other capital-related costs, making it a very expensive endeavor to undertake. Once up and running, the plastic bottles are able to achieve a 50-50 split between recycled and virgin material, which is better than before but still not what we would consider to be a closed-loop model. As a result of this large-scale operation in South Africa, the country has one of the highest recycling rates in the developed world at 48 percent, with most other nations four times lower or even with rates of zero. Because this method of mechanical recycling is very capital intensive, it is a centralized supply chain model of collecting and processing plastic waste and therefore bears high operational costs that make it cost prohibitive. Other methods to achieve higher recycling and reuse rates include Coca Cola's goal of 50 percent recycled content in its bottles by 2030 through a greater use of plant resins, a misapplication that does not address this polymerization problem and therefore is more of a marketing ploy than anything.[3] The problem has less to do with the input material than in how these materials are polymerized, de-polymerized, and re-polymerized.

We have got to face reality that while recycling is a useful practice, it is cannot solve the problem of Peak Plastic. Tactics such as mechanical recycling, the shortcut process of not depolymerizing, can never lead to an economical and sustainable solution that it is thought to be; the yield rates are too limited, the material can only be recycled once before its structural integrity declines, and therefore it can never lead to a sustainable model. We have got to think about this problem as one of polymerization/de-polymerization/re-polymerization, a full supply chain solution and not just one link in the chain, as occurs today. A greater understanding must be provided to consumers to know what happens after it lands in that bin, to know that this is not the end of the problem, but rather the beginning. With conventional recycling as a practice becoming more of a religion, based more on faith than knowledge, it becomes difficult to provide a clear picture of truth and consequences. Rather than a false narrative of recycling being good or bad, it needs to be understood for what it is, and what it is not.

The Short-Term Answer: Polymerize and De-polymerize (Re-polymerize later)

How do we fix this polymerization problem? First, everyone should recycle as a starting point in order to mitigate this existing problem, at the least. Given the size and extensiveness of the plastic supply chain system, where it should be, and where it should not be (e.g., our oceans), and the sheer fact that its design, manufacturing, and distribution grew so rapidly without an adequate closed-loop approach, there are no magic solutions to this polymer problem that can be implemented quickly to sidestep this problem. We need to implement a short-term solution that does more than today's conventional recycling program, but not yet ready for a complete *polymerization/de-polymerization/re-polymerization* that will effectively close the loop as a complete supply chain process. This interim step process that I'm calling for will take steps forward in closing the loop of plastic, which is good, but not get to the end state closed-loop system. This will be achieved through a supply chain system designed to enable the use of chemical recycling, the conversion process that can turn plastic waste back into its original monomers, the aim being putting the monomers back together again at a higher rate of reuse than is the case with mechanical recycling, when the polymer remains intact. Despite it being the superior method of recycling, chemical recycling is rarely used today because it is too expensive as a function of today's existing supply chain design. To fix this, we just need to turn this problem *inside out:* instead of determining that the more effective manner of recycling (chemical) is too expensive given today's supply chain system, let's find a way to modify our current supply chain system to make this technical method viable from both a cost and material yield standpoint, leading to it being a transformational option. This step forward, with the aid of a new approach to the plastic supply chain, will focus on the use of chemical recycling, or the breakdown of the components of plastic into its original elements, a true sustainable, closed-loop system. In this model, the original elements, or monomers, can be reused over and again, almost to near efficiency, the recycled material that is almost completely indistinguishable from the initial virgin polymer's molecules or components. The challenge left remaining to fully close the loop, and to enable a product to product conversion and reuse process is in being able to re-polymerize these materials back into the same material as before without the use of virgin materials. This is why this is only part of the closed loop; the last step to re-polymerize is the ability to profitably restore those monomers back into the same configuration. From my research, I believe an effective process to make this cost effective will soon be possible in order to truly have the first complete closed-loop system.

Taking this first step from only polymerizing to polymerizing *and* de-polymerizing will be a big step forward, including the assistance of the supply chain system to make it a profitable venture. Some techniques used today in chemical recycling are "solvolysis," or the breaking up of the polymers through the use of solvents, and "pyrolysis," the breaking up of the polymer through heat in the absence of oxygen.[4] Of course, neither of these methods are as natural as how nature handles it at an ambient temperature and pressure level in the conversion process, and are not complementary to be performed at a large-scale commercial setting, leading to only mechanical and large-scale solutions as the only viable option.[5] Researchers have been successful in achieving the technical benefits of chemical recycling in a purely laboratory setting of closer to a 100 percent reuse, with the greatest challenge being the high temperature and pressure requirements in the conversion process that is expensive and difficult to perform in a commercial setting, making a large scale operation, similar to the size of today's mechanical recycling plants, impossible.[5] Because my solution turns the problem inside out, it will address the supply chain challenge as the priority, making chemical recycling a viable technical option, and as a result, making a closed-loop system for plastic viable in the future.

This immediate approach to polymerization and de-polymerization is not the silver bullet that will solve Peak Plastic immediately, but it is the first step toward the ultimate design and supply chain solution for plastics. Using the chemical recycling process that is proposed in this solution, and changes to the supply chain system that I will address in Solution 4, true recycling rates could be as high as 75–80 percent, much higher than today's 15–30 percent, and much closer to a closed-loop system. Would the public be willing to divert away from the conventional recycling programs that most Americans believe already work in favor of this new model? Would the petrochemical industry be willing to consider a new model away from an endless stream of single use plastic, given the potential impact to its revenue and profitability? These questions will need to be answered, as well as how the overall supply chain system will work. Today's large industrial polymerization facility for plastic resin will produce approximately 100,000 to 300,000 tons of resin annually, and at best, a mechanical recycling facility will produce 7,000 to 14,000 tons in a year, a scale much smaller and not able to be synergistic in today's supply chain model. Any reuse of plastic material, including my design, cannot match the scale of these massive virgin polymerization facilities through tankers and railcars and ship out in similar economies of scale. Therefore, it is not just the technical polymerization process that needs to be reformed, making it more like the natural process of nature; the overall supply chain process must be repurposed as well, including a paradigm shift of what it means for an efficient design. In the past and present, an efficient design is a linear path of scale, and in the future, efficiency will be a circular design that is profitable but where industry and nature are in a closed-loop system.

Therefore, not only is the problem not a recycling but a polymerization problem, but it is a supply chain problem as well; how can these monomers that are the by-product of a de-polymerization process at high yields (97 percent) be reintroduced (reused) in a commercial supply chain system? Our solution will sell these depolymerized materials on the open market profitably as commercial grade monomers until a fully closed-loop system can put in place the full re-polymerization of the depolymerized material. In the future, there will be a process that enables recycled plastic back into a polymerization stage of the esterification process, pre-polycondensation, and then polymerization using a closed-loop system. Given the capital intensive model of plastic production today, it is unlikely, to say the least, that a petrochemical company would be willing to modify its highly efficient supply chain system currently in place to utilize depolymerized recycled plastic at a much smaller scale than its feedstock are taken in today. Therefore, in today's supply chain system, the loop is truly incapable of closing, especially since the petrochemical industry needs to rely more on plastic production than it has in the past, as I articulated earlier in the book. But, this solution will take the necessary steps to improve sustainability and progress closer to the end state!

Another consideration in polymerization is the use of these additives in the process that are worse threats than the base polymers themselves. How can we re-design and fulfill plastic designs to be nontoxic to us and the environment? Even if we depolymerize these plastics, before we officially are able to close the loop with new designs we must account for how to develop so many creative new materials in a nontoxic manner. Short-term, the best way to prevent these toxins from entering the environment is to depolymerize as much plastic as is possible because these methods offer the highest yield that leads to the most effective process. The good news is that this approach to de-polymerization will move us in the right direction, but what is really needed is the next approach to polymerization, one more consistent to that of nature: less toxic, more reusable, and able to be manufactured and reused in less dramatic conditions, not requiring such high temperature, pressure, and chemical additives.

The Ultimate Solution: Polymers Better Than Nature

To summarize: Most plastic produced today is just thrown away after one use. Some plastic is recycled for reuse but mostly via mechanical recycling techniques that merely grinds a polymer in order to reuse it, leading to low yield and reuse rates. As a short-term replacement, I am advocating for a de-polymerization (chemical recycling) of the polymers in order to achieve higher yields, as high as 97 percent, as the first stage toward a closed-loop system for plastic. A technical system for de-polymerization that can do so profitably and at high yields is an excellent first step, and the goal will be to find suitable market opportunities for resale of these monomers until they can use for a like-for-like process to be restored back into its original polymer resin. There will be market opportunities for these materials, but it will not be to large plastic resin manufacturers who have differently scaled supply chains as noted above. Not only do we need future materialization and polymerization strategies as good as nature, meaning high reuse rates of materials under safe processes, we need better systems than nature, given the scale of our production use. Today, the array of plastic designs required so outweigh what nature can provide that simply growing more plastic cannot be the solution. The future solution needs to achieve both the mass scale and complexity of today's use of plastic in a closed-loop system, like described above, no easy task. Yet when completed will be the holy grail, the ultimate solution that will solve Peak Plastic.

This time, the design for plastic must be one in a closed-loop system of polymerization, de-polymerization, and re-polymerization all wrapped up in a profitable supply chain system. For this to occur, we must reverse engineer the solution, as follows:

1. Plastic at its end of life needs to have a yield/reusability rate of over 95 percent with nontoxic and non-evasive by-products.

2. While in use, plastic needs to be nontoxic/non-evasive to the environment and humans. If littered, it needs to be bio- or photodegradable at best, or neutral to the environment at worst.
3. The polymerization, de-polymerization, and re-polymerization process needs to be nontoxic/non-evasive in its by-product and enable a closed-loop supply chain system.
4. The material and supply chain need to be designed in order to enable all of these while at the same time offering the same functionality and economics as today.
5. This future supply chain must have the potential to improve on nature rather than do less harm to it.

For this to be possible, the plastic polymer must not be designed solely from a material-science standpoint but rather an end-to-end supply chain system standpoint. In nature, polymers have been designed not as a function of itself but of an entire ecosystem; as a result, these materials were designed to have a positive impact to the environment in its composition and decomposition, even in being the building blocks and computer code of life in proteins and DNA. Perhaps in similar fashion as DNA, these new polymers of the 21st century will be able to be programmed in a manner where they are natural and non-evasive as well as bonded and unbounded as necessary to close the loop between an industrial society and nature. Imagine the redesign of today's PVC pipe to be strong and long-lasting enough to resist nature for 50–75 years: it is needed for the housing industry without leeching anything negative to users or the environment, and then at the end of its useful life, it could be depolymerized in order to be re-polymerized back into something else useful or even a new water pipe. Therefore, it's important to understand that this challenge of sustainability is not in the recycling of a material but in this conversion process of polymerization, de-polymerization, and then back through re-polymerization. As a result, and through means that are nontoxic and non-evasive, we will have closed the loop on plastic!

This means we need a new strategy for sustainability and design. The forces of nature that enable materials to polymerize, de-polymerize, and then re-polymerize must be redesigned in a new supply chain paradigm, avoiding the conventional approach of heating and high pressure to force-form these materials unnaturally to have all the benefits of being nature proof while in its afterlife being able to switch to being broken down in nature. This means that polymers must not be designed in a lab in totality, but rather in a supply chain, one that includes our environment as an extension to it as well. A polymer scientist should be just that, a scientist of polymers of nature and industry and how they become synonymous in some way, in a supply chain, with each other. How can our polymers rid themselves of extreme heat, pressure, and toxins within a market setting? I think this will be possible through emerging processes and technologies, such as material genome design, supercomputing, and 3-D printing, plus with the perspective of the overall supply chain, these designs should be possible in less than a decade, which will be a transformational paradigm shift for the plastics industry. When this happens, and it will, the forced programs of recycling will dissipate and the fate of the synthetic plastics industry will evolve or go extinct. What will become of it is a full-fledged 21st-century approach to materials, one where man becomes God-like not in his ability to control nature and society but rather to enhance both in a manner that was only possible in the past through the laws of nature. When this day occurs, and I think it will be in less than a decade, it will be interesting to see whether the most famous biologists will continue to see us as lumbering robots or blind machines raging toward our own destruction.

And yet, who can blame anyone for putting their chips down to bet on their cynical views with these biologists rather than believing this new approach to material science where supply chains merge with natural chains in the next decade? I would not blame you, but just as Dustin Hoffman's character in the 1967 movie *The Graduate* was given advice, I'm going to give it to you today: "There's a great future in plastics. Think about it. Will you think about it?" If you think about it, you must think about it differently, and plastic must become a much different material than it has become up to the present. While today it may be difficult to define as a material, in the future it will become easier to do so; similar to the polymer of nature, but much more diverse: as stretchy as polyester, as crunchy as Styrofoam, and able to become over 50 percent of the materials required for an airplane to fly across the planet. And then, thereafter, it will be able to be depolymerized for reuse at best and, at worse, harmlessly thrown into a forest to be reused by nature. It will be defined by a much more sustainable closed-loop system of polymerization, re-polymerization, and de-polymerization that matches nature: one of low energy, low pressure, low toxicity, and near 100 percent reuse. How will this be possible for a material to be protective of nature and then become a part of it, therefore being inherently better than nature? This will be the big transformation, the bringing together of everything that has been good from our circular environmental systems brought together with the best of our linear, one-way industrial ones. It becomes a way, the only way, that a population of 8 billion people on the planet can grow to 10 billion and persevere rather than just barreling toward Peak Plastic as a result.

Indeed, as I stated in the beginning of this chapter, polymers are the story of life, and nature's approach to polymerization is the impetus for us and our ability to redesign our world in a manner that hasn't been possible by any other species. Today, we must atone for our actions, first by closing the linear process of one-way use of plastic by making the best of poorly designed material, then by redesigning the material, and then creating a supply chain that is better than nature itself. Why should we not think this is possible, given the possibilities of artificial intelligence, artificial body parts that make our existing body parts less constraining, and renewable energy that is not only better than today's fossil fuels but as good as the sun? It will be a world where our technology will not only be in sync with our natural presences but improves on them as well. With this being the case, shouldn't our technology and supply chains be able to make polymers much better than nature?

References

1. Polymer Solutions News Team. (January 28, 2014). Green and natural polymers are on the rise. Polymer Solutions. Found at: https://www.polymersolutions.com/blog/green-and-natural-polymers-on-the-rise/ (accessed June 28, 2018).
2. BIO Intelligence Service. (2013). Study on an increased mechanical recycling target for plastics. Final report prepared for Plastics Recyclers Europe.
3. McCoy, Kevin. (January 19, 2018). It's the real (clean) thing: Coca-Cola wants 100 percent can and bottle recycling by 2030. *USA Today.* Found at: https://www.usatoday.com/story/money/2018/01/19/its-real-clean-thing-coca-cola-wants-100-can-and-bottle-recycling-2030/1047248001/ (accessed June 28, 2018).
4. Al-Sabagh, A. M., F. Z. Yehia, Gh. Eshaq, A. M. Rabie, and A. E. ElMetwally. (2016). Greener routes for recycling of polyethylene terephthalate. *Egyptian Journal of Petroleum* 25 (1): 53–64.
5. Hopewell, J., R. Dvorak, and E. Kosior. (2009). Plastics recycling: Challenges and opportunities. *Philosophical Transactions—Royal Society of London Series B Biological Sciences* 1526: 2115–2126.

CHAPTER NINE

Solution 4: A Closed-Loop System for Plastic

Now It's Possible!

Close the loop, closing the loop, a closed-loop system: there are many similar terms to describe what has been to this point only a metaphor rather than a practical application. Environmentalists, academics, public-policy analysts, and the like have used these terms to demonstrate how if plastic is just thrown into a bin and then returned somewhere (unknown) a circular, closed-loop system will be possible. Just Google the term "closed-loop recycling system," and you'll likely get some simple concept of how the by-product of one process or product becomes used for making another product. Sounds simple, doesn't it? The plastic being recovered from The Ocean Cleanup Project is hypothesized, according to its founder, to be turned into tomorrow's sunglasses and Nike shoes, as if it becomes a closed-loop system. While undoubtedly, this will happen, it is by no means in fulfilling a closed-loop system rather just a tactic. The outflow of plastic from all of the world's production sources in today's linear supply chain system is much greater than the amalgamation of all of these present-day tactics to recapture and reintroduce back into products; some of you may consider this to be a cynical statement, but it's just the math. The only, I mean *only*, way to for a closed-loop system of materials to begin to close for plastic is what I discussed in Solution 3: through the process of polymerization, de-polymerization, and re-polymerization in a profitable manner, which does not exist today. Simply capturing, cleaning, grinding, and reforming plastic into commoditized products sold at a very high price solely for niche consumers who will pay a premium to feel better about the oceans is not a closed-loop solution. Just as is the case in nature, there are no shortcuts, such as to simply recycle the materials (whatever this means) or use organic feedstock or whatever. Because today's plastic and its supply chain system has not been designed to fit in a closed-loop system (addressed with Solution 2), the starting point is a partial closed-loop system of polymerization and de-polymerization, as was addressed in the last chapter with Solution 3. Or maybe the first step forward must be to acknowledge a problem in the first place; let's stop fooling ourselves that real closed-loop systems are possible today through more recycling bins and environmental-conscious citizens and government policies. So step one is to acknowledge that we have a major problem to correct through today's system of collection and mechanical recycling; any legitimate form of a closed-loop system would not lead to less than a 7 percent recycling rate worldwide. Next, we take a major step forward by partially closing the loop by using the superior technique of chemical recycling and a transformational approach to supply chains that will be discussed in this chapter and lead to a 70–75 percent recycling rate, assisting in the prevention of Peak Plastic. Building upon this, the final step is to make all plastic optimal to both our synthetic and organic worlds. As I have noted frequently in this book, this will only be possible through supply chain systems in order to transform our materials to exceed and be an extension of nature. Even with the benefit of advanced processes and technologies, this will only be possible through an understanding that this is the path and must be the focus. There is no doubt that aligning all of the present-day stakeholders involved—from producers, suppliers, retailers, consumers, and even the environmentalists, all having their own beliefs and self-interests—into one path forward that focuses on the supply chain system to become integrated with nature in a profitable manner will be a challenge to achieve in a timeline required to address Peak Plastic. Therefore, this is why information and education is so important in shining a light on this problem of Peak Plastic and relate it to why there is no closed-loop system will be important because the statistics do not lie; perhaps the greatest problem in creating a closed-loop system is not necessarily a definition of what it is but rather what it is not.

The Partial Closed Loop: A Community-Based Supply Chain System

This new definition of a supply chain system must drive innovation and sustainability as much as it drives efficiency, profitability, and scale—a paradigm shift from the present that both can be achieved concurrently. The best approach to achieve this joint objective is through a major change in the supply chain to a community-based supply chain system, one that localizes the problem and opportunity rather than solving it by defining efficiency only as a matter of scale, capable of providing products for a large region, nation, or world as the primary progression goal. Think about this strategy in the context of the plastic bottle: it originates today from natural gas or petroleum, an industry that must rely on scale and efficiency as well as geographical locations across the planet. From there, resins are produced using an enormous amount of energy and other chemicals, requiring massive scale, efficiency, and safety to be effective, also specialized only in certain regions around the world. Next, this very low-cost commodity of a plastic bottle is filled with another commodity, water, and sent to a retailer to be sold for a dollar, enabling sufficient profit to be had by everyone in the supply chain; the source of this water is also often regional, based on scalability and quality standards of the water source. Another example of this *plastic strategy* is the production and distribution of the cheap plastic toy in your child's McDonald's Happy Meal that must be made in China or some other far-off region in the world through an optimization (as we call it) of inexpensive feedstock, energy, labor, and logistics. In these models, the *optimization* is calculated from start to consumer use, but not all the way through a closed loop, as the reverse logistics of returning the used commodity material back to where it was produced would blow up the financials, no longer making it rational to do so. This is why much of the plastic products that you consume and use are made in regional and global supply chains, making it impossible to close the loop, especially since China stopped taking on our used plastic trash.

Rather than transforming the supply chain through a community-based approach that I am proposing in this chapter, too many policy analysts have advocated a bottle bill program of some sort, although only for certain types of plastics, in an attempt to create an artificial monetization of the plastic material through mandating a deposit be added to the cost that the consumer has to bear, but this is backward thinking. How is the supply chain loop closed once this happens? In places like California and Sweden, both known for being progressive in these programs, the economics have never achieved what the policy analysts have designed it to be; the system will lead to a collection of the materials, pay the consumer back his or her deposit, pay the waste collector some processing cost, but somewhere, there is someone left holding the bag (of trash) when nobody wants it after the process is over. I've asked the question to many of these proponents: How can it ever work if nobody wants the plastic at the end of the loop? The true purpose of this process economically is for the material to be reused and turned back into a like-for-like valued commodity, such as a bottle-to-bottle process. These bottle bills are often just a forced effort of monetization that fails when the 5-cent deposit is swallowed up through administrative costs of the program rather than a real value of the plastic to the community. Certainly I applaud any effort that seeks to solve this plastic problem, but it must be done in a manner that understands supply chains and how they create and/or reduce the value of a material in the process.

In this community-based supply chain model that I am proposing, the plastic gains monetary value at the locality rather than being shipped back into the large national or regional supply chain, as is often the case with bottle bills wherein the supply chain system is left unchanged. Why does this community-based supply chain system increase the value of this used plastic material when the other approaches do not? For one, in plastic's present state, we must acknowledge that it was never designed to be reused, and therefore, in its waste state, it offers little reuse value. Therefore, in a community-based supply chain system, the plastic is being collected and processed within the same community where it became waste so no value is being lost in the logistics (processing, shipping, storage) of these materials to a regional, national, or even international location, such as China—that would suck all the value from the commodity material in logistics costs. By holding the line on the value of this material at the onset, the community retains any potential for the material to be monetized during the technical conversion process, which is the point where value can be achieved. At this point, using a community-based technical process for chemical recycling, the used plastic is then efficiently processed and converted into valuable chemical monomers using the chemical recycling process of de-polymerization, noted in Solution 3; my proposed approach to chemical recycling can achieve yields of over 95 percent, and therefore if collection and adoption rates are sufficiently high, the recycling rate can easily achieve the 75–80 percent target profitably, as noted in the last chapter. As I also noted in the last chapter, this community-based supply chain system must be modified to enable the technical solution to be economically possible to solve this plastic problem as the only viable approach in this partial closed-loop model. Seeking to recycle and reuse plastic waste materials through regional, national, and even global schemes—such as shipping plastic waste back to China—is nothing more than a mitigation attempt to reduce waste, and not a very effective one. Once a great source for plastic waste from around the world, China is reducing the amount of plastic that it is willing to receive from the rest of the world, creating a crisis to the existing recycled plastic market. A community-based supply chain solution will be the first-of-its-kind recycling and supply chain system for plastic that achieves a high recycling/reuse rate without government intervention and runs as a profitable venture on its own, without gimmicks such as sending useless plastic 4,000 miles away to somewhere else! Perhaps most importantly, it enables value at the locality, providing incentives for the municipality and its citizens to do something constructive with the plastic waste.

Another key to this new approach to a sustainable supply chain is to establish secondary markets for the processed monomers that are an output from the chemical recycling process. Recycled PET from a mechanical recycling process can be profitable and useful in niche markets but cannot be a sustainable option for the overall supply chain market if our goal is to recycle *all plastic* and achieve higher yield rates than what exists today, which is less than 50 percent. In contrast, monomers restored in the process from the community-based supply chain should have the same marketability as virgin monomers from a technical specification perspective, with the exception that they are being produced in small-scale community-based operations rather than mega-scaled petrochemical production facilities. For example, from a chemical recycling perspective that we are developing in our research for PET plastic, the process can generate monomers of ethylene glycol (MEG), terephthalic acid (TPA), and sodium sulfate, all materials with value on the open commercial market and profitable at current market prices. The challenge is that these materials at a local scale cannot be reintroduced back into the material streams of today's mega–supply chains even though we believe them to be the same high grade of chemicals produced from virgin fossil fuels; this can be addressed through markets with a connection to the local community. Rather than appealing to governments to step in, this is where the community gets involved because they can profit from it: it can use its purchasing and municipal influence for companies to do the right thing relating to investing in the local economy and addressing the environmental problem. Therefore, it isn't an environmental solution through putting a bottle bill tax on a product, nor some other form of subsidy or mandate, but rather by modifying the supply chain to connect it to the local community and environment. Despite what is often conjured when a term like "community" is involved, this is about private enterprise and a market-based approach that takes into account more of the variables rather than restricting it to just a few, as occurs more often today with today's mega–supply chain. In general, it requires a new way of thinking, something that is absolutely necessary given the immaturity of our existing system in regard to its stewardship to us as citizens and to the environment.

Community-Based Supply Chain for the World's Slums

Now, let's look at how this system can work today in places where traditional supply chain and markets do not. Take, for instance, Namibia, the small nation in Africa that was discussed in chapter 3, a place that has struggled geopolitically over the past few hundred years and now is on the rebound after receiving its independence from South Africa in 1990. As a result Namibia is one of the youngest nations in the world but is still carrying burdens from its past, having been subjected to what it often called "the first Holocaust" when it was a colony of Germany, and then afterward in being a colony of Great Britain and South Africa, facing an apartheid system of great economic and social inequality for many decades. Even though it is making great strides on the path of independence, 22.4 percent of

Namibians are living below the poverty line, and 37.4 percent of those are living in rural areas, making it one of the most unequal societies in the world.[1] With its burgeoning middle class that seeks consumer goods without sufficient process-manufacturing or waste-management systems, this developing nation not only must receive many imports from its neighbors, but it is saddled with the waste—a major detriment in this regional supply chain model. Given its lack of a structural supply chain system in its own country, and the persistent problem of drought that has affected over half of its geography for a prolonged period, Namibia relies on bottled water from South Africa. This of course leads to a disastrous material stream of a significant amount of PET plastic that enters the nation as a finished product without a viable path to anything of value after use, in turn leading to an ecological catastrophe. From a conventional linear supply chain definition that is improperly designed to address such environmental challenges, the system is working exactly how it is supposed to: product demand of bottled water moving from large manufacturing operations in South Africa linearly to Namibia for consumption. However, in a nation like Namibia with significant poverty challenges, this often manifests into a cycle of poverty and environmental damage with local waste pickers working below the poverty line to ineffectively sweep up the waste. This model is all too familiar in developing nations around the world, as its growing number of consumers demand goods that local supply chain systems cannot deliver, requiring regional supply chain systems from outside of their own countries to fulfill, which is a financial detriment to the economy. In many cities and rural communities, a lack of clean drinking water also leads to bottled water taking the place of inadequate municipal water systems, trading the problem of a lack of clean water with the ecological crisis of plastic waste. This problem becomes its own closed-loop system, but one of economic despair and ecological damage.

In a developing nation like Namibia that needs bottled water for its population given a lack of manufacturing, municipal water, and waste-management infrastructure, implementing a community-based closed-loop system has the potential to not only eliminate single use, one-way plastic waste through supply chain and technological innovation, it has the opportunity to reduce poverty by monetizing this waste in an economy that desperately needs value-added approaches for the economy and environment. In Namibia's capital city of Windhoek, there is a contrast between some of the richest areas of the nation and the poorest; in the northern region of the city is one of the largest slums in Africa, called Katutura, translated from the Herero language as "the place where nobody wants to live." In this slum, a typical waste scavenger will collect enough PET plastic bottles to fill five 5-kilogram bags, or 25 kilograms a day. At a fetch rate of N$0.39 a kilogram for bottles, or less than US$0.03 a kilogram, this translates into a daily wage of US$0.74, which is significantly below the latest United Nations definition of world poverty of US$1.95. Collecting over 800 PET bottles in a day (831) could take an entire day, yet does not enable the individual to escape severe poverty conditions as a result of his or her effort. In the neighboring nation of South Africa, there is a general rule of N$2.00 per kilo (approximately US$0.15 per kilo), which is five times higher than Namibia given its supply chain but still a woeful way of making a living. As you can see from this example, a supply chain system that is viable for the collection of plastic bottles to be mechanically recycled can lead to some economic improvement, but not much, and it only leads to a 50 percent recycling rate and small gains in taking people out of poverty, though this is much better than the extreme poverty found in such nations as Namibia without any local recycling plant at all.

In Namibia today, PET plastic bottles are collected in households for shipment to South Africa for reuse and thrown into dustbins that are sent to landfills to be picked through by waste scavengers or thrown into the natural environment, leading to environmental degradation. Presently, the most viable model is for these used beverage containers (UBC) to be exported to South Africa even though it is being done so at very low purchase prices; it is the best approach suitable for waste mitigation yet doesn't offer much, if any, economic benefit to the Namibian economy. This becomes a significant burden to Namibia's recycling business, and in particular the waste pickers who are at the bottom of its food chain, given many problems such as high logistics costs back to South Africa, regulations in South Africa, and low commodity prices that restrict participation by almost everyone other than the poorest in society and offer little economic incentive for the development of collection centers.[2] As is the case in many developing nations without proper recycling policies and infrastructure, it is easier to dump these containers than it is to recycle them, and this can unfortunately lead to these used containers eventually getting inadvertently dumped in bodies of water and other natural habitats.

According to my research, this community-based, closed-loop supply chain system will lead to at least a fivefold increase in fetch rates to waste scavengers in Namibia, bringing them from being well below the official UN poverty rate to twice as high. This favorable outcome is possible because it reduces logistics costs of shipping used plastic bottles to South Africa below the value of the material and uses a new technological innovation to convert these waste materials into market-grade chemicals that are needed in the local Namibian economy. In turn, the higher fetch rates received by the waste pickers stabilizes the economic value of the plastic material in the local economy through the creation of a multiplier effect of waste being turned into economic activity instead of crime and environmental damage. This in turn could lead to the enablement of other process-manufacturing operations in Namibia, such as paints, textiles, and abrasives, to name a few. From an environmental perspective, Namibia's low plastic bottle recycling rate of less than 10 percent, with what *is* collected being sent to South Africa for its betterment and not Namibia's, will increase sevenfold to over 70–80 percent, and the proceeds from the process will able to be reinvested back into the local community.

This community-based supply chain system is significantly smaller in scale by design in comparison to a large mechanical-recycling operation in comparison; this system would likely start as a 1,000-annual-ton local community-based supply chain system rather than a 7,000–14,000 ton annual operation at scale. But this is the point of the local community-based effort; rather than achieving scale at the detriment of the economic activity of the community where consumer products come in from long distances and waste remains, let the local supply chain turn the waste back into value. As a result of this, the degradation of the waste worker in these slums, relegated to the role of a waste scavenger, is transformed into one of a waste contractor, a difference that is more than just a name change. The waste contractor is provided a collection bin, registered into the system, and provided requirements to participate in the program. By doing so, we will provide dignity to the role and, as a result of a higher fetch price, require improved material collection specifications that will reduce recycling pretreatment costs. UBC collected today in developing nations achieves less than 50 percent of the yields in developing nations due to the contamination of the material.

By increasing the fetch rates of plastic waste to multiples of its present value, there will be greater demand for it as a remanufacturing feedstock, and a true *closing of the loop* through market principles not mandates. While some plastics will be more viable at the start given an ability to profitably chemically recycle them (PET plastic versus PE and PP, which will be more of a challenge), eventually this program can be implemented for all these materials. Through this method of monetizing the collection of used plastic locally in order to offer economic opportunity in developing nations, as well as lower processing costs (less than US$0.05 a container), high material yields from the conversion process, and an output of high grade chemicals, a community-based closed-loop supply chain system can be implemented that is a transformational approach to collection that raises the living standards of waste pickers in the developing world while reducing waste.

Solutions for U.S. Inner Cities

The United States may be one of the most prosperous nations in the world, but it isn't without areas of significant economic despair. The water crisis in Flint Michigan that happened in 2014, and still exists to a certain extent, is an example of how America, an economic superpower, has cracks in its infrastructure that is dangerous to the health and welfare of certain communities, often to those who can least absorb such challenges. In some of these areas, there are water pipes over 75–100 years old in desperate need of repair due to the threat of bursting and lead exposure. In 2001, the American Water Works Foundation predicted a crisis to occur in the future if these issues weren't addressed, and they have not been since that report over 16 years ago.[3] Estimates to fix these challenges nationwide range from $384 billion by the EPA to over a trillion dollars over 25 years (if commenced immediately) by the American Water Works Association.[3] Finally, according to the National Resource Defense Council, if waterlines continue to be replaced at their current rate, it will take 200 years to correct the present-state infrastructure challenges, and then, of course, we would be in arrears for other repairs.[4] This water crisis has led to a health crisis in Flint and possibly the culprit of a Legionnaires' disease outbreak that killed 10 people and affected 77. Currently, the water quality in Flint has been deemed as returning to safe levels, but since all the lead piping won't be replaced until 2020, residents are being told to continue to use bottled water for their potable needs, although the state government has recently ceased its bottled-water program.

In these types of circumstances where municipal water systems are unable to serve local communities, there will be a greater dependence on commercial bottled water, as is the case in developing nations that do not have sufficient systems and standards. Ironically, many of today's large bottled-water manufacturers rely on approximately 50 percent of its water in the United States sourced from a stable municipal water supply, meaning the consumer will need to buy water from a commercial provider rather than getting it from the public utility! In a town that is 10 miles from Flint, Michigan, Nestlé Waters produces 3.5 million bottles of water a day from a public resource for $200 a year in paperwork costs because individuals or companies in the United States do not have to pay for water by volume if they extract it themselves; meanwhile, residents of Flint pay some of the highest prices in the nation for their contaminated water.[5] This public policy to private enterprise paradox is a contributing factor to the economic insecurity and increase in plastic waste, causing a ripple effect in the economy, the environment, and our health.

The model I am developing for in the United States is a similar community-based, closed-loop system for bottled water and plastic recycling for the benefits of these depressed areas such as Flint, Michigan. Rather than facing the health challenges of an antiquated municipal water system and a bottled-water supply chain system that profits from publicly sourced water, a community-based supply chain system can enable a self-reliant enterprise through a small, nano-factory operation for water filtration, packaging, and plastic bottle recycling. As a result of modifying how the supply chain system works regarding the inflow of bottled water using the plastic waste as a feedstock resource for a chemical recycling operation, the community can profit from waste rather than be burdened by it, therefore breaking this negative closed loop of environmental and economic degradation. The operation is small enough to be able to be managed by the community, and it can scale by setting up other community-based systems in a hub-and-spoke model rather than a monolithic mega-factory at an economy of scale and capital investment level that would never be proposed for an inner city. The financials work in a different manner than a traditional supply chain system, but they do work as a commercial application rather than a nonprofit. As a result, inner cities have an opportunity to grow economic enterprises for the residents of its community while the plastic waste that is litter today can become a driver for growth.

Imagine how a community-based supply chain solution could transform lives and our view of plastic waste in a depressed economic area like Flint, Michigan; Baltimore, Maryland; or rural counties in Kentucky and West Virginia, parts of our nation that have been abandoned of hope. Today's supply chains find little opportunity to invest in these communities, and as a result, their only purpose is as pockets of consumption, making the supply chain work in one direction while leaving the plastic waste behind. Even if there are technological improvements made to change how plastic is recycled, it would not help these areas where environmental challenges are most pressing. But what if these areas could jumpstart a change in the supply chain system to make otherwise impractical methods of recycling, like chemical recycling, more viable? What if today's waste that cannot be profitably recycled in today's supply chain could become a feedstock resource and a multiplier of economic growth in tomorrow's supply chain? Areas without sufficient integration to our regional and national supply chain systems, like Flint, Baltimore, and Appalachia, could become hubs of turning waste into commercial products while at the same time improving the environment. It's an idea that could solve multiple problems through its supply chain system, creating a new mode of how we design, produce, consume, and reuse.

Fully Closing the Loop

These community-based supply chain systems are a major step toward closing the loop in these distressed areas of the developing and developed world, and they can be implemented across the planet. I predict that these systems could easily increase the plastic recycling rates on aggregate from about 7 percent to much higher than this, perhaps quadrupling to almost 30 percent in less than five years. Of course, this is great news and a significant improvement to today's conventional policies but also not enough to address the problem. What will be done to address the other 70 percent of the problem in order to have a fully closed-loop system? Hopefully, at the same time that Solution 4 is being put into place as a community-based supply chain system, Solutions 2 and 3 are happening to enable material and product and supply chain designs for more livable materials (Solution 2) through a more radical approach to polymerization (Solution 3). As this occurs, the supply chain system grows in scale and complexity, transforming itself to be capable of leading to higher recycling/reuse rates. Through better culture (information and transparency), design, technology, and supply chains, markets will be disrupted through these new plastic supply chains, finding better, cheaper ways of servicing the customer. So it appears as if a community-based supply chain system is antithetical to all that we have been taught as best practices for supply chains and markets and the starting point or incubator for this new approach of a closed-loop system.

As a separate matter from the topic of this book, the future of supply chain management will be a peer-to-peer (P2P) economy that will turn many things local, or maybe the best term is "glocal" (global and local). Consumers will be producers and producers will be consumers, and through a combination of automation, 3-D printing, and digital design (e.g., material genome), as well as block chain technology to create trusted peer-to-peer transactions, linear and anonymous supply chains that lack transparency will be a thing of the past. Through technology and transaction, materials can be local, as well as being localized; plastics will need not be made from feedstock from tens of millions of years ago, dug from deep holes and processed only in massive industrial sites that can withstand high temperature/pressure operations. The term "economy of scale" in a linear supply chain system will be a relic of the past, just as a tree isn't manufactured locally from the best seedlings from Europe and soils from South America, but rather within its own ecosystem. Localized supply chains in some communities have either been ignored or collapsed, such as the examples of Namibia and Flint, Michigan, and must become "open for business," including as a node for the larger system rather than vice versa where these embattled communities are in a state of disrepair that mutually degrades the economy and the environment in a concurrent manner. In the Flint example, the community's disconnection to any sort of supply chain system, other than as a consumer, had occurred long before its water system failed through the changes that have happened in the global automotive supply chain system. Therefore, the water crisis of 2014 is just a perpetuation of the existing cycle of poverty and environmental despair where its citizens do not have access to clean water, and its local citizens must purchase bottled water from a private enterprise that essentially gets free water from the state, the same state that cannot provide clean water to the Flint community despite the high prices they are paying? Today's conventional supply chain model will not organically self-organize into these local, community-based systems by its means of practices due to its model of efficiency built upon the "bigger is better" mantra by definition. Yet when it comes to a need to provide a solution for an inner-city community without clean water and massive amounts of plastic waste, this supply chain model is the better option, and furthermore, can be done in a manner that promotes local, profitable entrepreneurship rather than public funding that is often ridiculed by the private sector as the problem. Given the challenges in our 21st century society, it seems rather disjointed to consider our most rational approach to supply chains is a linear, one-way supply chain system that pumps out waste that is bad for many local economies and the environment, and that a lack of a real closed-loop system is better for either the economy or the environment. For some stakeholders, such as the petrochemical industry, bottled water companies and even retailers do benefit from this linear, single-use proliferation of plastic waste, fostered by cheap plastic, but in no way should they be the scapegoats, and neither should we as consumers. Through changing the paradigm, we should unleash the innovators who understand that the challenge is a supply chain one not a material-science or public-policy matter. Hopefully, this book can create a spark to those groups of innovators who can get started on working on the partial and fully closed-loop systems that will bring an end to this futile banter of whether the economy is more important than the environment, and vice versa.

Why not unleash a new version of supply chain management where new designs of polymers are possible to be polymerized, de-polymerized, and re-polymerized into a closed-loop system of near 100 percent yield efficiency that is profitable, affordable, and nontoxic to us and the environment? Through Solution 2, an open-source, open-access approach to designing a new approach to plastic, and Solution 3, an understanding of the problem with plastic to address, a new form of supply chain management can close the loop in a market-based model that can provide novel products to consumers without destroying the environment. The model could be transformed as follows: first, a group of innovators will reverse engineer existing plastic resins, such as PVC and polystyrene, possessing the same technical specifications that exist today but nontoxic and able to be locked, unlocked, and relocked again (Solution 3) as necessary given the time in the product life cycle. After this radical approach to design using today's technology and the technologies of the future, the manufacturing of these plastic resins can be fulfilled in a 3-D printing facility locally rather than being produced globally and shipped long distances, which effectively negates any possibility of economic viability to be recycled. Because these manufacturing facilities can be profitable as local operations, a closed-loop system will not only reduce the potential to leak plastic waste into the environment, there will be a powerful pull of this material back to the manufacturing facilities in contrast to the ineffective push model that exists today of mitigating plastic waste. Of course, the elephant in the room is the mega-scaled, highly profitable petrochemical industry and its monopoly of all the plastic products that we use as well as its viability of an economic engine for local, national, and global economies. But it will be up to the consumer to decide the model that makes the most sense to them relative to all the factors involved.

Whether the focus is on poverty and a lack of clean water in Africa where developmental aid has failed or a water crisis and recycling programs in the United States where public policy has failed, it's time for the supply chain to drive the change because it is often the source for why change needs to be driven in the first place. The traditional supply chain models of scale, such as the oil and gas and petrochemical industries, are a different definition of efficiency that must be transformed by us as consumers in markets rather than by governments. These dominant-design supply chains will continue to do what they do best, and we need to allow them to do so but not as our only option for how to meet the challenges of our society, economy, and environment. These companies and the corresponding supply chains that have been successful for so many decades, particularly the oil and gas industry that still lays claim to trillions of dollars of resources still in the ground able to be utilized, will not transform to new models easily. But it will be outside of the traditional supply chain model where new models of innovation will take root to solve for new problems, such as the need for a closed-loop system for local economies and sustainability practices. How it all plays out will be up to us as consumers and citizens, with the probability of a need for both the traditional model of supply chain and novel definitions to address the complexities that exist, such as those mentioned in this book.

References

1. Namibia National Planning Commission. (2015). Poverty and deprivation in Namibia 2015. Windhoek: National Planning Commission.
2. Schenck, Catherina, Derick Blaauw, and Kotie Viljoen. (2012). Unrecognized waste management experts: Challenges and opportunities for small business development and decent job creation in the waste sector in the free state. Pretoria: International Labour Organization. Found at: http://www.greengrowthknowledge.org/sites/default/files/downloads/resource/Unrecognized_waste_management_experts_ILO.pdf (accessed June 29, 2018).
3. Ferris, Sarah, and Peter Sullivan. (April 25, 2016). Clean water crisis threatens U.S. *The Hill*. Found at: http://thehill.com/policy/energy-environment/277269-a-nation-over-troubled-water (accessed June 29, 2018).
4. Olson, Erik, and Kristi Pullen Fedinick. (June 2016). What's in your water? Flint and beyond. National Resource Defense Council. Found at: https://

www.nrdc.org/sites/default/files/whats-in-your-water-flint-beyond-report.pdf (accessed June 29, 2018).

5. Glenzain, Jessica. (September 29, 2017). Nestlé pays $200 a year to bottle water near Flint—where water is undrinkable. *The Guardian*. Found at: https://www.theguardian.com/us-news/2017/sep/29/nestle-pays-200-a-year-to-bottle-water-near-flint-where-water-is-undrinkable (accessed June 29, 2018).

CHAPTER TEN

Solution 5: Fixing the Invisibility Problem

Solving the Visible, and the Invisible?

Before I present this last of the five solutions, I would like to recap the first four to frame the big picture of how to address Peak Plastic: First, the current-state hemorrhaging of plastic into the environment needs to be controlled through the use of short-term solutions to pressure resin manufacturers through markets to disclose the toxic additives in their products, and to provide ample information and transparency to consumers in the developed world and education to the developing world of this problem of plastic. Also, we need to control the amount of plastics going into the ocean and our bodies through the banning of primary micro-plastics, and enable the informal waste picker network to control the outflow of plastic waste that is concentrated in the top polluting nations and rivers. Hopefully, undertaking these short-term solutions will give us some breathing room to start with Solution 2, which is open-sourced and open-accessed plastic via a 21st-century innovation and market model. Through this solution, plastic resins can be designed for every stage of its supply chain life cycle, including reuse. This leads us to Solution 3, which is the full technical process for a closed-loop system that ensures polymerization can lead to de-polymerization and re-polymerization so that today's plastic product can be efficiently, sustainably, and profitably looped back into a like-type product. As a result of a more efficient polymerization process, this is the first step toward the closed-loop supply chain (Solution 4) as a partial closed-loop system that polymerizes and depolymerizes into monomers to be resold. For the complete closed loop to be enabled, a newly designed plastic polymer is designed for a closed-loop system later on in the process.

Solutions 1–4 will lead to a complete transformation for the future of our planet and ourselves without affecting us as consumers. Hopefully, the day will come in the year 2030 when all plastic produced will be in a closed-loop system, leaving little waste and no adverse impact on the environment and ourselves, but it will take the full 12 years between today and the year 2030, the target date of Peak Plastic. Yet according to the *Future of the Sea* report produced for the UK government, the total amount of plastic debris in the ocean will increase from 50 million metric tons in 2015 to 150 million by 2025, meaning our challenges will continue to grow even as we put effective measures in place.[1] This means that by the time our first four solutions are in place, it is quite possible there could be four times more plastic in our oceans and other natural environments than today. Worst of all is the problem of invisibility, an inability to see the real impact of plastic in the environment due to the limitations with our current technology. If we can't see it, not only can we not clean it up entirely, but we may also be underestimating the real impact on us and the environment today as a result. If we achieve Solutions 1–4, we will have made significantly more progress in combatting this plastic problem than has occurred over the past decades but still be woefully short in solving the problem. To counter what remains, we need to address this invisibility problem.

The invisibility problem may be a greater challenge than we believe or can understand, maybe even leading to my own estimates of Peak Plastic to be understated. While there has been some attention focused on the plastic that we can observe in the environment, and even that we are beginning to have better measurability through new technologies, insufficient attention is being paid to nano-plastics, or materials smaller than 100 nanometers. As a point of comparison, a human hair is approximately 60,000–80,000 nanometers wide. These nano-plastic materials are so small that they can likely pass through biological membranes, impacting processes that include blood cells and photosynthesis.[2] These nanomaterials are difficult to discover on landmass and are almost impossible to detect through our waste management systems; as a result, these nearly invisible particulates end up in natural water systems such as oceans and even our drinking water through a natural fragmentation process that occurs in ocean and river banks, as has been simulated in research, as well as ultraviolet radiation and microbes. On land, there are numerous opportunities to create nano-plastic particles, including the drying of clothes, plastic manufacturing, fertilizers, 3-D printing, biomedical applications such as drug delivery, and so many others that are probably impossible to catalog.[3]

Who really understands the magnitude of this problem? Nobody. In a recent 2017 UN study, it was noted that while plastic in the ocean undoubtedly has an impact on marine life, it is unclear its impact on our own human health; while larger pieces of plastic will likely not be absorbed by a fish, these smaller nanoparticles probably are absorbed, and the toxicology reports are lacking.[4] There is much we do not know of what is in the ocean and where; plastics lighter than polypropylene will float and eventually gather in the gyres, while a significant amount of plastic is heavier, meaning that it will sink to the ocean floor and become a part of the food chain.[5] If these particulates can be absorbed into a fish's body, including the toxins attached to the nano-plastic particles, they will certainly become a part of our bodies as well when we ingest seafood and other sources of nano-plastics, such as our municipal and commercial bottled water. Once these nanoparticles are in our bodies, are we able to excrete them or do they integrate into our systems, our cells? Our knowledge is improving, but it is certainly not keeping pace with our plastic production and use. This is unclear and even with Solutions 1–4 as noted in this book, we must be skeptical.

Solution 5 must address this invisibility problem that exists, but to what extent it exists we do not know. How can we feel sufficiently confident that a containment of plastics through design, polymerization processes, and a closed-loop system allows for us to contain this influence even if future plastic is not synthetic or toxic? How confident are we that the nature of the leeching process of plastic over the decades, without sufficient testing and technology, is not leading to a change that we do not understand? For certain, our vision of the problem is poor, but how poor? If 83 percent of the world's population has drinking water contaminated with plastic, with the United States as the highest at 94 percent of its samples contaminated, as well as beer, honey, sugar, salt, and practically everything we ingest, what will we learn when our tools are better?[6] Is it time today to gain a greater focus, or shall we wait?

Solving Peak Plastic and Still a Problem?

This topic reminds me of a sad story that a relative of mine encountered in his battle with lung cancer. According to the story, the doctor broke the news to my relative when he asked, "But why do I have cancer? I stopped smoking 25 years ago?" The doctor replied, "Yes, but you still smoked." Even if we implement Solutions 1–4 successfully over the next decade, is it possible that our detection of the problem is too crude to have made enough of a difference, given the damage already done? Can our radical solutions and others, such as these floating contraptions on the ocean implemented by The Ocean Cleanup project, take enough of the plastic from the oceans, and if we deployed these devices across the Earth, would it be enough, undetected, or too little, too late? We have to ask the question given the damage that has already been done whether we have done enough, and whether our means of detection can tell us the truth. If anything, it's more likely that I am understating rather than overstating the problem of Peak Plastic.

Wouldn't this be a shame if these solutions are not enough, and we are not able to restore the natural environment back to its original conditions before these nano-toxins emerged and became so plentiful? The first step is greater awareness through a campaign that our knowledge needs to improve to what we cannot see; the field of nano-toxicology is emerging, and there needs to be a focus on how synthetic materials, particularly plastics, are impacting our health. For Solution 5, there will be a push for greater research in the field of nano-toxicology and the impact of synthetic plastics to the environment. Without more research, we could be inadvertently ignoring a bigger problem than exists today simply because we lack the knowledge and tools to understand exactly what we are looking for. There are literally hundreds of trillions of plastic micro- and nanoparticles littered across the land and sea, some that are hosting toxic materials, and this is the final mile of Peak Plastic that must become a part of public awareness. Given the compounding nature of our use of plastic today, simply turning off the faucet is not sufficient; we must understand what is spilled, what it means to us, and how to clean it up.

Technology to the Rescue

Technology needs to come to the rescue and help us understand what we cannot see, but how is this possible given approximately only 5 percent of the ocean has been discovered?[7] Today, nanoparticles can be seen through the use of high-powered instruments, such as electron microscopes, but this is only practical in lab and not field settings, and especially not in such remote areas of the ocean that have yet to be discovered. For us to understand the real impact of what lies beneath our sights in every corner of the planet, including places never seen before, we need portable, powerful devices to detect and then access in the cleanup effort. Certainly, The Ocean Cleanup is a solution, but it is too limited in scope and technical capability with the potential to address only a sliver of the overall project to the best of our understanding. This undoubtedly seems to be the most difficult, if not impossible task to undertake, to find all these synthetic, toxic nanoparticles underground, underwater, in the deepest trenches, the highest peaks, and places on Earth, places that are largely undiscovered. Is this not the last mile of the journey of Peak Plastic? If our solutions work as promised, Solution 1 will stop the bleeding long enough for Solution 2 to create the overall design; Solution 3 will deliver the polymerization process that is akin to and then better than nature, and then Solution 4 will first begin to close the hole of the closed-loop system before shutting it altogether. And then we will feel great about ourselves in finally achieving a solution to plastic that is in sync between the economy and the environment without understanding what cleanup is left to finish the job. As difficult as Solution 5 may be, it is the last mile, the final journey, in restoring the planet back to where it was before we designed a material that was foreign to nature. I believe if awareness is brought to the attention of micro- and nano-plastics, researchers will find solutions and the true final cleanup stages will take place.

Now, all five solutions have been proposed and are ready for implementation. It's a long journey of what must be done to eliminate Peak Plastic, but aren't these steps to identify the problem, measure, analyze, improve, and then sustain it? Yes, and with a problem as complex as that of plastic, this is certainly a challenging task. I think the problem is very clear even if measuring its impacts are not; whether we are absolutely clear of the dangers of plastic to us and our environment, we do understand that it is a synthetic material that is out of place in a growing proportion in our bodies and environment. This we know for sure, even if we do not know yet what it means. In this book, I've promoted what I believe to be a nonbiased view of plastic as either good or bad, but rather both, and neither at the same time, and no judgment upon conventional solutions such as government policy and technology other than it being insufficient. Now it's time to act, to frame up the problem to the extent of its scale and scope in order to respond accordingly with solutions. It'll be up to our rationality and imagination as to whether we believe we can continue to ignore what is now among us everywhere and anywhere or jump at an opportunity for a 21st-century supply chain system and level of care for our environment.

References

1. Chow, Lorraine. (March 21, 2018). Ocean plastic projected to triple within seven years. *EcoWatch*. Found at: https://www.ecowatch.com/oceans-plastic-pollution-2550762861.html?utm_source=EcoWatch+List&utm_campaign=a50c6d8f00-EMAIL_CAMPAIGN&utm_medium=email&utm_term=0_49c7d43dc9-a50c6d8f00-86120461 (accessed June 29, 2018).
2. Yong, Chin W. (2015). Study of interactions between polymer nanoparticles and cell membranes at atomistic levels. *Philosophical Transactions of the Royal Society B: Biological Sciences,* 370 (1661).
3. Wright, Stephanie L., and Frank J. Kelly. (May 22, 2017). Plastic and human health: A micro issue? *Environmental Science and Technology* 51 (12): 6634–6647.
4. Yildirimer, Lara, Nguyen Thanh, Marilena Loizidou, and Alexander Seifalian. (2011). Toxicology and clinical potential of nanoparticles. *Nano Today* 6 (6): 585–607.
5. Boucher, J., and D. Friot. (2017). Primary microplastics in the oceans: A global evaluation of sources. Gland, Switzerland: IUCN, 43.
6. Carrington, Damian. (September 5, 2017). Plastic fibres found in tap water around the world, study reveals. *The Guardian*. Found at: https://www

.theguardian.com/environment/2017/sep/06/plastic-fibres-found-tap-water-around-world-study-reveals (accessed June 29, 2018).

7. NOAA National Ocean Service. How much of the ocean have we explored? Found at: https://oceanservice.noaa.gov/facts/exploration.html (accessed June 29, 2018).

CHAPTER ELEVEN

Summary: Make It Happen!

Who Should Do What?

Martin Luther King Jr., John F. Kennedy, and Mahatma Gandhi—all heroes from my childhood—lived by the following quote from Gandhi: "Action expresses priorities." This quote perfectly illustrates my closing question: Is Peak Plastic a priority in public policy? According to a 2017 Gallup poll, while more than half (55 percent) of Americans ranked the environment as a top issue, it was significantly lower in importance than terrorism threats (76 percent) and the economy (73 percent), and while 75 percent say they are concerned about the environment, only 20 percent say they act on that concern all the time.[1] Yet what is the more likely event to happen any day of your life, for your body to be subjected, in some unknown manner, by the influence of plastic in our modern-day society or a terrorist attack? And what is the best manner to improve the economy: a backward-looking approach to manufacturing and supply chain systems, or one moving forward? This lack of immediate concern over the impact of plastic on our lives seems to be more over a lack of information, not just regarding the potential hazards but in the opportunities as well. Even if future research studies found this unusual amount of synthetic material surrounding us as not harmful to any substantive degree, which would be a wonderfully shocking finding, the economic opportunity to improve the health of our oceans, and to link the industrial to natural supply chain systems would be a great win-win for everyone. Hopefully in this book, I've painted a picture of an equal amount of concern and abundance of opportunity to motivate us toward action rather than alarm. As such, it makes no sense for us to not act.

My theory is that there is no resident expert on it, so words are difficult to use to describe the impact of plastic on us and the environment, and yet there needs to be more effort on this topic despite the challenges. In taking this journey of writing the book, what has surprised me the most is how vast the topic is, plastic as a noun, verb, and adjective. The attempt to discuss plastic as a generic *one-size-fits-all* material that can and should be recycled is just too simplistic; it is becoming downright dangerous in preventing a serious dialogue on how to address its impact on our oceans and bodies, and as well, the economy if we choose not to do anything about it. I do not consider myself to be an expert on plastic, and I doubt that anyone else should consider themselves in this matter either; I was willing to take on the challenge of writing this book in order to stake the claim that none of us are experts on a topic so vast and difficult to understand. Of course, taking on a broad topic to consider the polymer science, economic, supply chain, environmental, cultural, and health aspects of such a topic will lead to critiques from experts in each of these individual categories as to how certain explanations or depictions were insufficient, incorrect, or unnecessary. This book is a strawman of sorts in the hopes of an end-to-end discussion of plastic as an opportunity to build a platform on a topic so complex to address the industrial and environmental setting that exists today; any obsession regardless too refined or a level of details at this stage is to be missing the point. Each of these aspects is opportunities for the future and is complicit to the problem at the same time. This love-hate affair with plastic is a surefire way to a path of inaction out of discourse and a lack of information and knowledge that must be addressed immediately. Everyone should focus on the bigger picture, which is the path of doing nothing or simply mitigating the problem should be past us; the time to expect policy analysts and those clinging to decades of a failed understanding that the industrial supply chain does not need to be in sync with the natural one must be left behind. Today's sheer scale and complexity problem of nearly 400 million tons of plastic annually, strewn across the planet, is a different scenario than when it was a material used by fewer of us in a smaller circle in fewer permutations of products, and as a result, we need as much of a transformative approach to solve it. And like it or not, it has already pushed its way, advertently or inadvertently, into the top priorities of the developing world given its environmental crisis, in growing concern of our food and drink all over the world, and as just a general concern given the high level of concentration of this synthetic material in everywhere and anywhere, just as a matter of common sense. As such, it makes no sense for us to not act.

Hopefully, the case that I have made is that it is more than just a coincidence of the impact of a greater material flow of plastic in our industrial and natural supply chains to some of these environmental, health, and safety concerns; just as with the cases of cigarettes and opioids, a clear cause-and-effect relationship may not possible to conclude, and it is up to us in the absence of such a difficult bar to clear as to whether this should lead to us just ignoring the relationship or digging in to learn more about it through deeper research. There's certainly no turning back from the successes of the 20th-century supply chain that continues into the 21st century of how today's global system works and will continue to grow. Nobody reading this book should expect this mega-scaled, uber-efficient supply chain system to voluntarily tear down their models of growth in favor of smaller, circular closed-loop systems and material bans and nor should any of us want this to happen. The solutions to these problems will therefore likely arrive from innovators working much differently than in the past, working in more of an open-sourced, open-accessed model where private profitability is incented and possible but in more of a balanced model than how it happens today. After we put in place some Band-Aids to get control over what's happening, this combination of a new model of innovation, design, material science, and supply chain can bear fruit, connecting our industrial system to that of nature, leading to a *design for good* rather than simply mitigating damage. Manmade methods of polymerization using high temperature and pressure must change, and while the new methods may never be as efficient as nature that has transpired over millions of years, it will get better just by aspiring to be more like it rather than careless of these natural processes. Polymer chains will be able to successfully lock, unlock, and relock as materials making today's products long lasting and viable. This will first commence through a partial closed-loop system of polymerization to de-polymerization of the existing design to recycle via high-grade commercial monomers and then through a newly redesigned plastic that allows for the closed-loop system to occur with high yields and no toxic or invasive impact to the environment. Finally, our last task is to clean up the substantial mess that we will have made by 2030 to include the visible plastic waste as well as the more challenging micro- and nano-waste that is difficult to detect with today's process and tools.

The greatest challenge we face in this plastic problem, the problem of Peak Plastic, is the first step, the acknowledgment that this journey we need to undertake is not as much of an indictment on plastic as a little understood synthetic material in high use than it is a desire to better understand it to make it more practical and harmonious in our lives and ecosystems. But this will only be possible through engaging new actors as a part of the conversation, making the platform more about incentives to achieve profitability and sustainability rather than an economic or environmental priority as a false narrative. Of course, the economy always wins, and should always win: as I mentioned earlier in the book, economic growth has pulled literally billions in the world out of poverty, and we should not stop that progress now. The difference is that these solutions to Peak Plastic will be led by innovators, enabled by the supply chain system, supported by scientists and government policy makers, and able to drive both economic and environmental activity. But key to all this—and the only thing I am asking from consumers and citizens—is a paradigm shift in thinking; apathy is not an option because an informed public is the key to the problem always!

The first priority is a public awareness campaign to improve our knowledge of what plastic is without any fear that doing so will take things away from our consumer lifestyle if we follow these five solutions. This public awareness campaign should build upon the true thought leaders out there who already exist and are trying to educate the public on what's happening, such as Marcus Eriksen of the 5 Gyres Institute and the Plastic Pollution Coalition, among many other notable academicians and industry leaders. Their roles will be very critical, albeit slightly different, to bringing greater awareness and galvanizing support collectively rather than separately as well as building more support in order to welcome innovators, entrepreneurs, and investors to the party who can make the greatest difference in fixing the problem more so than mitigating or simply talking about it. And yet it will be key to build a platform for conversation as soon as possible, and as broadly as possible, in order to get started on this massively complex problem. Setting forth a goal of 2030 seems challenging to the task and requires that we get going on it as soon as possible.

The time is now to get started. The doors should be swung wide open for the innovators through private- and public-enterprise incentives—as it should also be for product designers, supply-chain professionals, and funders—to begin to work together to solve this problem in an open-sourced, open-accessed manner. These will be the real stars of the show, and those of us who are the advocates must aspire to recruit the best and the brightest to innovate and fund for us, to bring new players to the table. Our job as the old guys is to find new blood, interest, and money to build a platform of interest in how it is going to be solved from a new perspective. Without this step of galvanizing interest and new ways of thinking, we'll continue to be stuck in the present. With a focus on the Millennials who are interested in environmental matters, and especially if they can be handled in a profitable manner, we have a chance of transforming this linear, one-way supply chain system. In today's old-school model, there are concerns regarding different plastic formulas, such as PVC, polystyrene, and polycarbonate with BPA, but there are barriers of doing anything about it. Getting innovators and young advocates to work outside of the existing supply chain system or any government mandate on de-formulating these recipes will shed light where darkness exists today. Through research, public awareness will be funded that will lead to a multiplier effect, leading to new designs and supply chain models. Closed-loop design prototypes should begin to take form in about five years, and through an open-sourced and -accessed model, 3-D printers and distributors should begin to build mini supply chains of innovation. Given the exciting future of block chains, artificial intelligence, supercomputing, and material and genome design, these products should begin hitting the market in less than seven years, making synthetic, one-way plastic obsolete in less than 7 to 10 years. Is it possible that this new version of plastic can replace synthetic plastic so quickly as this model of innovation replaces the existing mega plastic supply chain that is almost a hundred years old itself? Despite the powerful market position of the petrochemical industry, this is possible and, dare I say, inevitable. The natural process of open-sourced, open-accessed markets is to enable capitalism, not to dissuade it, and a frictionless system through the use of local supply chains and 3-D printing will make it more practical in 10 years than we can imagine today. We just need to get past the first step for this to happen.

Not just gaining involvement from the innovators, product designers, and funders but from the science community as well. Where was the scientists' involvement in the plastic problem over the past 70–100 years? With a more open model, I am hoping that questions arise regarding our industrial approach to polymerization versus how nature performs its tasks; can our industrial approach of "heat, beat, and treat" be improved to something less severe, foreign to nature, and sometimes toxic? This is the key to change regardless of whether the feedstock is ancient hydrocarbons made from fossilized leaves or crops pulled from farms in the last month. If scientists find a way to polymerize a feedstock in a manner where it is nature-resistant when it needs to be, and then nature-friendly afterwards, it becomes a product that is effective in both the supply chain and the ecosystem. If not, can scientists find a material that can be polymerized and depolymerized in a closed-loop system so effectively that the material never leaves our supply chain system given its value, like aluminum? This is another discovery that must be achieved by science and technology, if there is the incentive to do so. Today, there does not appear to be sufficient knowledge and incentive for sufficient research capability and capacity to be put toward solving for it.

We will also need the best and brightest minds from my area of specialty, the supply chain management community. Rather than focusing on bigger, more single-minded determinations of efficiency, our industry needs to consider and perfect a new model of profitability that includes sustainability. In the future, supply chains will be more profitable when they are more multidimensional; community-based supply chain systems, for example, will redefine definitions of efficiency and profitability, opening up markets that are deemed as unprofitable today, such as developing nations and inner cities in the developed world. The legitimate links between poverty and environmental challenges have been a no-man's-land for markets as a function of inefficient supply chains, and turning our definition of the supply chain inside out as being localized will lead to new paradigm shifts. Not just in opening markets in Africa and U.S. inner cities but in the use of new technologies, such as block chains and 3-D printing that could work better in smaller, localized supply chains than mega, globalized ones. Such opportunities should draw young students to the supply chain profession as it becomes front and center in solving these significant matters in markets rather than public policy.

Lastly, the entire supply chain system must consider itself to have the same level of responsibility as a doctor has over a patient, where in this case we are the former and the environment is the latter. How have we gotten to the stage where we are today where there is an almost infinite amount of matter that cannot be seen yet is so dangerous to us and our environment? This needs to be the start of a better understanding of what is invisible to us and how our industrial world can never again create synthetic matter that is dispersed without controlling and knowing the consequences. To use the important saying from the medical field, *primum non nocere*, or "first, do no harm," we as product designers, material scientists, manufacturers, supply chain practitioners, and consumers must take a stance that our current-state consumption should not have a negative impact on our future generation's consumption ability, and without understanding today's plastic supply chain system, we cannot make this promise to them. Our knowledge of our plastic must be from beginning to the end of the process, and from the largest macro pieces of materials to the smallest, nearly invisible nano-plastic particles in order to get our arms around it as a 21st-century material. If not, we will always be in arrears, as our intention to use is greater than our intention to understand.

If We Stand Still, and Never Take the First Step

A book project like this to cross multiple platforms of research and understanding should be the first steps in defining a 21st-century problem of the magnitude of how we went from the one plastic bottle in design to something as complicated as Peak Plastic. Where do we go next, and what questions do we ask ourselves in this sort of debate? If you've engulfed yourself to the degree that I have regarding the usefulness and concerns regarding plastic and its use in the 21st century, you will no doubt see it everywhere and not be able to disconnect from it, both good and bad. Even beyond plastic as a material and culture, it can become all-encompassing if you possess the same high-powered microscope I have used in this project, which I'm convinced isn't nearly powerful enough. My only fear with this project is that I have left too much on the table and not provided a strong enough case to shift us out of neutral; I'm reminded of my last book tour when my audiences agreed on diagnosis, but not the remedies. Is it possible that you will agree, but are too overwhelmed with other concerns? Sure, information overload can often lead people to feeling as if there's no hope and thinking why bother doing anything about such an insurmountable problem. Yet if we step through the problems and the opportunities within one supply chain system, that of the connected industrial to the natural one, solutions made sense and seem possible when they have not before.

I am not adverse to different paths to take to solve these problems, but none of them should lead us back to where we are today. There is no turning back once you learn what is happening; as I have been writing this book, I have had one-on-one discussions with industry leaders, academics, and environmental advocates as well as group discussions that always seems to lead to the same question: "What did you just say?" So much of this seems so unbelievable that I am glad that I've written this book, but we must set forth and change the tide after we get past understanding that we are at this place we never thought we'd get to. Who among us old enough to think about what it was like to live in the 1970s or 1980s would have ever expected to be here, at this place, where we are debating such large percentages of unintentional plastic in our oceans, bottles of water, and even our own bodies? Journeying out on this problem and ending back where we started here today can be captured in the words of comedian George Carlin: "The Earth probably sees plastic as just another one of its children." That would be a strange place for us to end in the 21st century when none of us, or even our ancestors before us, would have ever expected the future to be a place that is tainted by a material so separate from our environment that it would take it over and change our lives as a result.

The Impossible Dream?

Hopefully, a problem this serious, yet without much conversation, can all of a sudden become a rallying cry to action. It may never become a top item for the twenty-four-hour news cycles that always seem to focus on the sensational and unimportant, but it must go viral in a 21st-century communications approach (e.g., social media) that drives the disruptive innovation to improve both the economy and the environment. What a concept, right? It will be up to nameless people who are innovators, entrepreneurs, and investors to drive this—not celebrities or others in high offices of power, but especially the younger generation of Millennials who are interested in balancing making of a difference while making a profit because these two are not contradicting objectives. I'm just hoping that this book will help raise awareness and become a catalyst for others to proceed down the path of the five solutions, which can be done without government involvement.

It's best to take on this challenge via 21st-century tools of open source, collaboration, closed loop, and integration with nature than a hostile takeover attempt over the dominant-design supply chain system of the petrochemical industry as some sort of government regulation as an evil to dismiss, which of course it is not, and we as consumers are its greatest advocates. So innovation and transformations will begin to reduce the use of virgin plastic polymers that fuels the waste, and as a result, more will be reused, leading to less plastic and toxins into our oceans and bodies in the meantime. The actions of innovators, entrepreneurs, and investors will catalyze consumers and vice versa, leading to change in the manner of our supply chain systems, and vice versa. Yet perhaps the greatest impetus to action will be from a push for greater trust and transparency in the consumer products that fill our homes, a push for consumers to understand this, and the suppliers to oblige. Isn't it clear that a lack of knowledge today is the main engine to apathy, and, eventually, a planetary crisis? Transparency and knowledge could be the flash point that links our industrial supply chain to nature, just as ignorance and a bit of knowledge, just like Dr. Frankenstein possessed, as what separated them. Perhaps the consumer, the Millennial consumers, will drive this in a manner never before seen regarding consumer passion and excitement.

Imagine if our industrial supply chains were in sync with nature: that the great potential of the 21st-century system that enables human progress could at the same time improve the nature of our environment. Imagine if our industrial supply chain systems were not only able to pull millions out of poverty but would also not debilitate those remaining in poverty to a worse circumstance, which is what happens today in those left behind, often as waste pickers and other misfortunes of indignity. Ironically, as more are pulled into the middle class, the pressure being placed on the environment makes it worse to those who are poor, creating a vicious cycle between poverty and environmental damage. Imagine if everything we did to the environment was able to improve upon it, if we could use technology and innovation to create materials better than nature, including a polymerization process that offered magnificent innovation and near 100 percent reusability. These are more than just grand illusions; they are becoming musts to figure out as the world's population grows. If we have learned anything from the devastating impact that plastic has had on the environment over the past recent decades, it is not what the worst of it can be but rather the best in what is possible with our knowledge and technologies of the 21st century.

Now, fast forward a century from today to the year 2118; what will the world think of our use of plastic in the year 2018? Will it be looked at as just the first steps of neophytes, as if we were toddlers, in understanding the usefulness of materials and supply chains in our modern society that we eventually outgrew and in the nick of time led to a world of balance between the industrial and natural, the synthetic and organic? Hopefully this is the case, and it won't be seen as the unwillingness of our generation to consider the resources of the world beyond just our immediate needs. If it is our generation that destroys the natural environment to the extent that economic and environmental sustainability is no longer possible, how will our descendants think about us? When viewed in this light, the seeming paradoxes or forks in the road between the environment and economy no longer are false narratives of choice; instead, we are left with no choice than to optimize both, and as soon as possible. It may seem to be a dramatic way to end this book, but it's the cold hard truth, and it's up to us to imagine the future as a function of today and to think about how we should act as a function of this future rather than just our immediate todays and tomorrows. What happened between the 1950s and today, 2018, should be no reflection as to how we view our world in 2058 related to what happened today. For our future generations, there can be no greater gift than that of a willingness to balance our present to their future, just as we should balance our industrial systems to be in sync with nature. If we think in these terms, nothing is too big of a problem, and our future will be brighter than the present!

Reference

1. Anderson, Monica. (April 20, 2017). For Earth Day, here's how Americans view environmental issues. *Pew Research.* Found at: http://www.pewresearch.org/fact-tank/2017/04/20/for-earth-day-heres-how-americans-view-environmental-issues/ (accessed June 29, 2018).

Index

Addition polymer, 24
Agriculture industry, 80
AkzoNobel "Imagine Chemistry," 103
Antimony, 23
Asia: largest polluters, 44; plastic impact, 9, 44–45, 50–52, 94–95

Bakelite, 16–17
Biology, 7, 15, 58–60
Bisphenol A, 29, 31–33, 36
Block chain technology, 91–93
Borlaug, Norman, 58
Braungart, Michael, 106
Butylated hydroxytoluene (BHT), 25

Cancer rates, 64
Center for Disease Control (CDC), 22
Chemical recycling, 21, 115–117
Chemicals, 37
China: aquaculture, 33, 52; plastic imports, 124; Yangtze River, 96
Cigarette epidemic, 61–63
Closed-loop system, 12, 121–122, 130–132
Coca Cola, 114
Condensation polymer, 24
Cracking, 15

De-polymerization, 112–117
Design for good, 104–106
Design revolution, 102–107
Di(2-ethylhexyl) phthalate (DEHP), 32
Dow Chemical, 17, 28
DuPont, 17

Endocrine disruption, 25
Environmental Protection Agency (EPA), 8, 22, 26–27, 88

5 Gyres Institute, 94, 144
Flint, Michigan, 23, 51, 128–130
Food and Drug Administration (FDA), 36, 109
Food concerns, 52
Formaldehyde, 29–30

Generative customization, 103–104
Geology, 15
Great Pacific Garbage Patch, 45, 49

High density polyethylene (HDPE), 24

Innovation, 58, 61, 67–70, 105–106, 144–146

Lorenz, Edward, 59–60, 143–144
Low density polyethylene (LDPE), 24

Materials Genome Initiative (MGI), 105–106
Mechanical recycling, 20–21, 112–113, 115–116

Microbeads/nano-plastics, 9, 50, 93–95, 136–137

Namibia, 46–47, 125–128
Nature's supply chain, 4, 42
Nutrition Labeling and Education Act of 1990 (NLEA), 109–110

Ocean: cost to clean, 95; ecosystem, 47–48; plastic impact, 48–50, 93–95; seafood danger, 50–52
Ocean Cleanup project, 44–45, 137–138
Ocean Conservancy, 95
Open-source capitalism, 99–102
Opioid epidemic, 62–63

Peak Plastic concept, 4, 6, 8, 37, 53
Peer-to-peer economy, 131
Petrochemical industry, 79–80, 112
Phthalates, 31–32
Plastic: additives, 31; base polymer, 19, 30; bio-feedstock, 5; bottle, 6, 22; classification, 19; definition, 6; deformulation, 107–108; health concerns, 21–22, 25–32, 34–35, 89–91; history, 16, 18; leeching, 17, 25, 31; manufacturing process, 16; miscellaneous, 29–30; production, 8, 18, 42–44, 80; tipping point, 4–5, 9; World War II impact, 4, 17–18, 41–42
Plastic Disclosure Project, 89
Plastic Pollution Coalition, 144
Plasticizers, 27, 31
Polycarbonate (PC), 29
Polyethylene (PE), 17, 23–24
Polyethylene terephthalate (PET), 19–23
Polymerization problem, 113
Polymers, 16–18, 30, 111–112
Polystyrene (PS), 17, 19, 28
Polyurethane, 29–30
Polyvinyl chloride (PVC), 16, 19, 25–27
Puerto Rico, 51
Pyrolysis, 115

Recycled PET (rPET), 20
Recycling: American's attitudes, 65, 90; band-aid, as a, 88–90; inadequacies, 26, 28–29, 46–47, 60, 63–64, 112; Namibia, 46–47; rates, 7, 19–21, 24, 42, 113–114, 122; *The Recycling Myth,* 75–78; South Africa, 46, 113–114; Sweden, 19, 27, 75–76
Re-polymerization, 112–117
Research limitations, 35–37, 61–65
Resin identification system (RIC), 7

Sheldrake, Rupert, 60
Solvolysis, 115
Supply chain: community-based, 124–130; definition, 7; ecosystem, as an, 81, 148–149; global challenge, 33–37, 52; innovation, 67–70, 81–83, 132–133, 143–146; linear model, 82; nature versus industrial, 42

Thermoplastics, 17–18
Thermosets, 17–18, 24, 29
Throwaway society, 52–53
Toxic Substance and Control Act (TSCA), 88
Tragedy of the Commons, 100–101

United Nations reports/programs, 44–46, 60, 136

Waste concept, 42–43
Waste pickers, 45–46, 95–96, 126–127
Water: contamination, 27, 32–34, 51, 137; infrastructure costs, 129; poverty and, 133

X-Prize Foundation, 103

Zero waste, 20

About the Author

Jack Buffington is responsible for warehousing and fulfillment for Miller-Coors, the second largest beer manufacturer in the United States. His professional background includes leadership positions in the fields of supply chain, operations, information technology, and finance across the consulting and consumer products sectors, both domestically and internationally. An adjunct and research professor at University College and the Daniels College of Business at the University of Denver, Buffington earned a PhD in Industrial Marketing/Supply Chain Management at the Luleå University of Technology in Luleå, Sweden. He is the author of several books, among them *The Recycling Myth* (Praeger, 2015).